The Expanding Universe

The Expanding Universe

Astronomy's 'Great Debate' 1900–1931

ROBERT W. SMITH

ASSISTANT KEEPER OF PHYSICAL SCIENCES,
MERSEYSIDE COUNTY MUSEUMS, LIVERPOOL

CAMBRIDGE UNIVERSITY PRESS

CAMBRIDGE
LONDON   NEW YORK   NEW ROCHELLE
MELBOURNE   SYDNEY

583072

Published by the Press Syndicate of the University of Cambridge
The Pitt Building, Trumpington Street, Cambridge CB2 1RP
32 East 57th Street, New York, NY 10022, USA
296 Beaconsfield Parade, Middle Park, Melbourne 3206, Australia

First published 1982

Printed in Great Britain
at the University Press, Cambridge

Library of Congress catalogue card number: 81–15471

British Library cataloguing in publication data
Smith, Robert W.
The expanding universe.
1. Astronomy—History—20th century
I. Title
520′.9′04   QB32
ISBN 0 521 23212 0

QB
981
.S69
1982

To my parents

That which is far off, and exceeding deep, who can find it out?

Ecclesiastes vii. 24

You will be interested to hear that I have found a Cepheid variable in the Andromeda Nebula.

Edwin Hubble to Harlow Shapley, 19 February 1924

Contents

Contents

Preface

One evening in 1917, a young American astronomer was applying the finishing touches to his doctoral thesis. A few years earlier he had given up a legal practice to pursue astronomy, and now his decision seemed to have been vindicated. The future certainly looked bright for him since he had been offered a post at Mount Wilson Observatory, probably the leading centre for astrophysical research. Moreover, the Observatory was soon to be equipped with what would be by far the most powerful telescope in the world, a 100-inch reflector, and one that offered the prospect of untold riches in terms of astronomical discoveries. Yet, with his thesis complete and this splendid vista before him, he sent a telegram to the Director of Mount Wilson announcing that he was off to join the American Army to go and fight in Europe. Six years later, this astronomer, Edwin P. Hubble, was to make a find with the 100-inch telescope that would dramatically end a debate that had lasted for centuries and that would all but finish the process begun by Copernicus in 1543 in removing the Earth from a privileged position in the Universe.

In late 1923, Hubble was examining photographs of the great Nebula in Andromeda. To his surprise, he noticed that a faint spot of light he had first identified as a nova seemed instead to be a star that regularly altered in brightness. Furthermore, the changes indicated the star to be a Cepheid, a type of variable whose distance could be easily calculated from the period of its light variations. With the aid of this momentous discovery and others that quickly followed in its wake, Hubble effectively ended the long-standing dispute on the nature of nebulae. The unambiguous message of the Cepheids was that very many of the hundreds of thousands of known nebulae were in fact 'island universes', star systems beyond our own stellar system, the Galaxy. Even more remarkably, within a few years Hubble, extending and exploiting the pioneering researches of many astronomers and mathematicians, was to provide what was soon to be seen as strong evidence that not only do island universes exist, but that their motions indicate that the Universe is not static, that it is indeed expanding.

The late 1910s and 1920s had also been a period in which the size of our own Galaxy (often referred to in the early years of the century as the 'Universe') had been drastically revised upwards. Hence when in 1932 the great English astronomer and physicist A. S. Eddington wrote his thrilling account of the latest developments in extragalactic astronomy in *The expanding universe*, astronomers had witnessed in less than a generation three sweeping changes in their view of the visible universe:

(1) the roughly ten-fold increase in the size of the Galaxy,
(2) the acceptance of the existence of external galaxies and
(3) the realisation that these galaxies disclose the expansion of the Universe.

In the chapters that follow I shall attempt to describe and explain these startling shifts. As all of them involve an expansion of some kind I have imitated Eddington and have termed this account *The expanding universe*. Its subtitle, *Astronomy's 'Great Debate' 1900–1931*, derives from one of the most famous events in astronomy in these years, the so-called 'Great Debate' in 1920 between Harlow Shapley and H. D. Curtis on 'The scale of the universe'.

The sources that I have employed to investigate these shifts are published works (principally scientific papers), the reminiscences of astronomers active in the early twentieth century or who knew astronomers themselves at work in that period and archive materials from the period such as letters and drafts of papers. The present study emphatically makes the point that we can hardly begin to understand the history of early twentieth century astronomy by a scrutiny of published materials alone. In particular, scientific papers, although efficient vehicles for transmitting the results of scientific inquiries, by their very nature obscure the actual activities of scientists. To penetrate the mask of 'public science', I have pursued other sources and have illuminated the published works in the light of these sources. The central concern of this book is with intellectual factors. This is certainly not to deny that, for example, sociological analyses of group loyalties and institutional patterns may well be essential to a full understanding of the intellectual shifts and I hope that this work will at least do something towards clearing the ground for future and very different studies.

Robert W. Smith
Liverpool, April 1981

Acknowledgements

Without the assistance of many librarians, archivists and astronomers, the examination of the sources on which this study relies would have been impossible. Particular thanks are due to the staff of: the Center for History of Physics at the American Institute of Physics, New York City; Cambridge University Library; Churchill College Library, Cambridge; the Henry E. Huntington Library, San Marino; the Institute of Astronomy, Cambridge; Lick Observatory; Liverpool City Libraries; Princeton University Library; the Royal Astronomical Society Library; the Royal Greenwich Observatory Library and Archives; Widner Library, Harvard. Also, Dr. Nils Hansson of Lund University, Dr. Kurt Pedersen of Aarhus University and Dr. Elly Dekker of Leiden University all very generously gave of their time to search through archive collections at my request. Of those astronomers who kindly gave me their personal recollections, I am especially grateful to Dr. C. D. Shane for his stimulating and patient correspondence. I must thank too Mrs. Shane for her always friendly and helpful advice on items in the Lick Observatory Archives.

  During the writing of this book (and the doctoral thesis on which it is based) I have benefited from the publications, criticism and advice of a great many people. That the account that follows is in many places less superficial than it would otherwise have been is due largely to the efforts of Dr. R. Berendzen, Dr. R. P. Cleaver, Dr. D. W. Dewhirst, Professor Owen Gingerich, Dr. R. C. Hart, Dr. Norriss Hetherington, Mr. David Howlett, Mr. W. G. Hoyt, Dr. M. J. Irwin, Dr. Barry Madore, Professor W. H. McCrea, Mr. Julian Ravest, Dr. John Schuster, Dr. D. Seeley, Professor Bengt Strömgren, Dr. and Mrs. Garry Taylor, Professor Charles A. Whitney and Professor G. J. Whitrow. To all of them, as well as numerous other friends and colleagues who have helped to clarify my thinking, I offer my very best thanks. I am also indebted to Mrs. Annette Woolford for performing miracles in turning my almost illegible scribbles into an accurate typescript.

It is a pleasure to acknowledge the help of the Twenty Seven Foundation who provided funds for me to travel to several archives in the United States. This visit was also aided by the fine hospitality of Professor and Mrs. Erwin N. Hiebert.

Above all, I owe a very special debt to Dr. Michael Hoskin for first suggesting the subject of this work to me, his own penetrating researches into early twentieth century astronomy, his unfailing encouragement and his invaluable editorial comments on various drafts of this book.

Abbreviations of manuscript sources

In the notes to chapters, manuscript sources are abbreviated in the following manner.

Aarhus	University of Aarhus, Denmark.
A.I.P.	Centre for History of Physics at the American Institute of Physics, New York City, U.S.A.
Allegheny	Allegheny Observatory Archives, U.S.A.
Hale	Microfilm copy of the G. E. Hale papers.
Harvard	Widner Library, Harvard University, U.S.A.
Huntington	Huntington Library, San Marino, U.S.A.
Leiden	University of Leiden, Netherlands.
Lick	Lick Observatory Archives, U.S.A.
Lowell	Lowell Observatory Archives, U.S.A.
Lund	Lund University Library, Lund, Sweden.
Princeton	Princeton University Library, Princeton, U.S.A.
R.A.S.	Royal Astronomical Society Archives, London, United Kingdom.

1

The revival of the island universe theory

Nebulae: star systems or nebulous matter?

By the middle of the nineteenth century the nature of the nebulae had been debated for over a hundred years and two main questions had emerged: first, were these dim patches of light remote star systems whose light merges to give a milky effect, or were they made of a luminous fluid; and second, if they were groups of stars, were these star systems part of, or associated with, the Galaxy, or were they themselves vast independent galaxies – 'island universes'? To many it seemed that the observations made in the 1840s with Lord Rosse's giant telescopes in Ireland – particularly those with the celebrated 72-inch reflector – had demonstrated that, given a telescope of sufficient light grasp, all of the nebulae could be resolved into stars. The claimed resolution of the Orion Nebula was especially persuasive.[1] J. P. Nichol, Professor of Astronomy at Glasgow in Scotland, had been a leading advocate of a theory that the nebulae are clouds of luminous fluid, but in 1848 he announced that the 'supposed distribution of a self-luminous fluid, in separate patches, through the Heavens, has beyond all doubt been proved fallacious by the most remarkable of telescopic achievements – the resolution of the great Nebula in Orion into a superb cluster of stars'.[2] Moreover, the resolution of numerous other nebulae into stars was taken as strong evidence for the island universe theory. This was because many astronomers, through a desire to adopt the simplest possible hypothesis, had taken up one of two extreme positions: *all* unresolved nebulae are systems of stars or *all* unresolved nebulae are masses of true nebulosity. For those who accepted the first alternative, then it was but a short and easy step to assert that the nebulae are indeed island universes: 'If it were the case that *all* nebulae were clusters of stars, some of them near enough for a powerful telescope to pick out the individual stars and so 'resolve' the nebula into stars, but others too distant to appear as more than a milky patch in the sky, then would it not be unreasonable to make *all* these star systems part of our own galaxy: a milky patch which was too distant for the individual stars to be detected and which yet was seen to occupy a sizeable portion of the sky must surely be enormous in size. In other words, to equate all nebulae with star clusters would be to lend strength to the conviction that other island universes or galaxies exist beyond and outside our own.'[3]

Not all were prepared to accept this reasoning. Some nebulae had long been suspected of displaying observable motions, and this suspicion was given further credence when in 1853 Liapunov, Director of the Kazan Observatory in Russia, detected changes of brightness and form in the Orion Nebula.[4] If the alterations were genuine, then the Nebula could not possibly be a remote cluster of stars since such a system could not display motions visible over a short time period. A more convincing example of variations in a supposed distant star system was 'Hind's wonderful nebula'.[5] In 1852 a faint, round nebula had been discovered by the British astronomer J. R. Hind. Many watched its gradual rise in luminosity, but in 1856 it started to fade and it had almost disappeared when Auwers searched for it in 1858. In October 1861 he found it had dimmed beyond sight, yet Otto Struve of Pulkowa Observatory could just observe it in late 1861 and by the following March he felt that it had brightened.[6] Here was apparently irrefutable proof that a nebula had undergone startling and rapid changes; hardly the expected behaviour of a huge star system. It now became necessary to take the other accounts of variable nebulae more seriously, and the island universe theory was thereby brought into question.

In 1864, the debate on external galaxies took a completely unexpected turn when William Huggins, a British practitioner of astronomical spectroscopy, aimed his telescope and attached spectroscope at a planetary nebula in Draco:

[I had a feeling] of excited suspense, mingled with a degree of awe, with which, after a few moments hesitation, I put my eye to the spectroscope. Was I not about to look into a secret place of creation?

I looked into the spectroscope. No spectrum such as I expected! A single bright line only! At first I suspected some displacement of the prism, and that I was looking at a reflection of the illuminated slit from one of its faces. This thought was scarcely more than momentary; then the true interpretation flashed upon me. The light of the nebula was monochromatic, and so, unlike any other light I had yet subjected to prismatic examination, could not be extended out to form a complete spectrum . . .

The riddle of the nebulae was solved. The answer, which had come to us in the light itself, read: Not an aggregation of stars, but a luminous gas. Stars after the order of our own sun, and of the brighter stars, would give a different spectrum; the light of this nebula had clearly been emitted by a luminous gas.[7]

Huggins had thus established in striking fashion the reality of gaseous nebulae, and, because of the assumed polarity of choice between the nebulae as star systems and the nebulae as clouds of gas, he argued against the existence of any island universes, even insisting that the purely continuous spectrum of the Andromeda Nebula that he had observed was due to gaseous matter under special conditions and *not* stars.[8]

The foundations of the island universe theory were further undermined by analyses of the distribution of nebulae on the sky. In 1864, Sir John Herschel listed the five thousand known nebulae in the *Catalogue of nebulae and clusters of stars*.[9] R. A. Proctor, a tireless British writer on astronomy, marked on charts of the sky the positions of the 4035 'irresolvable' nebulae in the *Catalogue*, and after completing this tedious task he was sure there was 'no mistaking the fact that a true zone of nebular dispersion . . . exists in the heavens'.[10] Furthermore, the plane of the Milky Way coincided with that of the zone. Now if the nebulae are truly independent of the Milky Way, it would be curious indeed for them to avoid just that region of the sky marked out by the Milky Way. Proctor thus contended that the nebulae form a part of our Galaxy.

During the 1880s two events occurred that seemed to have settled once and for all the issue of the existence of visible external galaxies. First, in 1885 a 'new star', or nova, blazed very near the centre of the Andromeda Nebula. The nova provoked great interest since the Nebula was one of the largest, and presumably nearest, of the nebulae.[11] Whatever the cause of the nova, it now seemed inconceivable that the Nebula could be composed of millions of stars: at its brightest the nova had attained a luminosity of roughly one-tenth of the entire Nebula, and how, astronomers asked, could a single star rival the gigantic light output of an immense stellar assembly? (The new star was in fact a supernova, but at the time there was no reason to believe in the occurrence of such violent stellar outbursts.)

In 1796 the brilliant French mathematician Pierre Simon Laplace had speculated that the planetary system had condensed out as a series of rings from a cloud of nebulous matter.[12] The plates of the Andromeda

Nebula secured in the late 1880s by the British astronomer Isaac Roberts (with a specially designed 20-inch reflector) lent force to Laplace's hypothesis. When some of these photographs were exhibited at a meeting of the Royal Astronomical Society in London in 1888 they created a sensation. Twenty-two years later the Savilian Professor of Astronomy at Oxford, H. H. Turner, vividly remembered the excitement of that evening: 'One heard ejaculations of "Saturn", "the Nebular Hypothesis

Fig 1. The Andromeda Nebula photographed in 1888 by Isaac Roberts.

made visible", and so on.'[13] Roberts's accompanying notes announced that 'those who accept the nebular hypothesis will be tempted to appeal to the constitution of this nebula for confirmation, if not for demonstration, of the hypothesis', since here one apparently can 'see a new solar system in process of condensation from the nebula – the central sun is now in the midst of nebulous matter ... The two [smaller] nebulae [associated with the Andromeda Nebula] seem as though they were already undergoing their transformation into planets'.[14]

By the late 1880s, then, an astronomer could exploit the bright line spectrum of some nebulae, the peculiar distribution of nebulae, the photographs of the Andromeda Nebula and the 1885 nova to form a seemingly overwhelming case against the island universes. Indeed, in 1890 the eminent historian–astronomer Miss Agnes Clerke declared:

No competent thinker, with the whole of the available evidence before him, can now, it is safe to say, maintain any single nebula to be a star system of co-ordinate rank with the Milky Way.[15]

Two years before its close, the nineteenth century's verdict on the island universe theory was passed by the writer of a history of astronomy who pronounced that the 'island universe theory of nebulae, partially abandoned by Herschel after 1791 ... but brought into credit again by Lord Rosse's discoveries ... scarcely survived the spectroscopic proof of the gaseous character of certain nebulae'.[16]

Despite the theory's almost complete fall from favour, it was nevertheless destined to undergo an extraordinary revival, a change in fortune that was to be due to observations made of a particular class of nebulae, the spiral nebulae.

Spiral nebulae and astrophysics

Mid-nineteenth century professional astronomers had been little concerned with the physical nature of the celestial bodies. Their energies had instead been directed to measuring, as precisely as possible, the positions and proper motions of these objects. Professional astronomers endeavoured, as the astrophysicist Samuel P. Langley stated in 1885, to determine '*where* any heavenly body is, and not *what* it is'.[17] Once a body's motion on the celestial sphere had been traced, the next task was to understand this motion, usually within the confines of the theory of universal gravitation. In 1888, the triumphs of this dialectic between data and theory had even seduced Simon Newcomb, the leading American positional astronomer, into claiming that while it 'would be too much to say with confidence that the age of great discoveries in any branch of science had passed by', yet 'so far as astronomy is concerned, it must be confessed that we do appear to be fast reaching the limits of our knowledge'.[18] But at the very time Newcomb was uttering his confident declaration a new discipline was struggling to establish itself, a discipline that would eventually ally itself with the traditional astronomy of position. This interloper was the so-called *New Astronomy*, the science of astrophysics.[19]

The two basic tools of astrophysics were the photographic plate and the spectroscope. With the aid of the photographic plate an astronomer could detect and permanently record sources of light too faint to be seen by the naked eye;[20] by use of the spectroscope the spectrum of an astronomical body could be scrutinised and, the astrophysicist hoped, its chemical composition inferred. Astrophysicists were thus permitted to tackle questions thought previously to be beyond the power of astronomy: What are the stars and nebulae made of? How does a star's chemical composition change through time? Is there a correlation between a star's composition and its motion? None of these questions could have been posed, let alone answered, with the concepts and techniques of the traditional astronomy.

The establishment of astrophysics was a long process, and one that was impeded by the entrenched attitudes of many of the positional astronomers. In 1899 William Huggins, one of the most influential pioneers of astrophysical research, recalled the reception in 1868 of his spectroscopic measurements of radial, or line-of-sight, velocities:

To pure astronomers the method came before its time, since they were then unfamiliar with the spectrum analysis, which lay completely outside the routine work of an observatory. It would be easy to mention the names of men well known to whom I was 'as a very lovely song of one that hath a pleasant voice'. They heard my words, but for a time they were very slow to avail themselves of this new power of research.[21]

Some were reluctant to accept the concepts and methods of astrophysics because the novel astrophysical techniques often gave very poor results. The American astronomer Asaph Hall wrote in 1866,

For one, I shall be glad to see improvements in methods of observing, but for a very large part of this accurate work of astronomy I don't yet see how photography is to help much ... It seems doubtful whether it is well to insert such a method between the observer and the result, since new sources of error are brought in.[22]

In consequence, the first generation of astronomical photographers and spectroscopists, men such as William Huggins, L. M. Rutherfurd and Henry Draper, generally lay outside of the professional astronomical community, and in 1885 Miss Clerke had emphasised that astrophysics 'is, in a special manner, the science of amateurs'.[23] But by 1910 astrophysics had become a recognised discipline with clearly formulated methods and concepts, defined programmes of research and solid institutional support; to a large extent astronomy and astrophysics had become intertwined into a common enterprise.

The main aim of astrophysicists was to understand the course of stellar evolution.[24] A result of this was that by the early years of the twentieth century one class of nebulae, the spiral nebulae, were being examined more closely than ever before because they were viewed as proto-stars. In 1898 the American astrophysicist James E. Keeler began photographing bright nebulae and clusters with the 36-inch Crossley Reflector of the Lick Observatory in California.[25] His early death two years later temporarily halted the programme, but in this short time he had estimated the total number of nebulae observable with the Crossley at around 120 000, an entire order of magnitude higher than the accepted figure of known nebulae. Furthermore, Keeler held that most possessed a spiral structure. This was an astonishing discovery because, although spiral nebulae had first been observed by Lord Rosse and his colleagues half a century earlier, only a few dozen, at most, had so far been found. The spiral nature of prodigious numbers of nebulae was now employed against the island universe theory. Miss Clerke, for example, was of the opinion:

[The] relationship between the various orders of nebulae is manifest. The tendency of all to assume spiral forms demonstrates, in itself, their close affinity; so that to admit some to membership of the sidereal system while excluding others would be a palpable absurdity. And since those of a gaseous constitution must be so admitted, the rest follow inevitably.[26]

That spirals belonged to the Galaxy was also an assumption of a new hypothesis that pictured proto-solar-systems as developing through a phase in which they displayed a spiral pattern. This hypothesis was advanced by the eminent geologist T. C. Chamberlin and his astronomer colleague at the University of Chicago, F. R. Moulton. In the first decade of the twentieth century the nebular hypothesis of Laplace that had long dominated other hypotheses of the formation of the solar system fell from favour[27] and by 1908 George E. Hale, the Director of

Mount Wilson Observatory in California, could write that 'it can hardly be denied that Laplace's idea of the development of the solar system must be reconstructed or abandoned. It remains to be seen what can be substituted for it.'[28] Chamberlin and Moulton's alternative, the *planetesimal hypothesis* (significantly called at the time by Moulton 'The spiral nebula hypothesis'[29]), had been launched in 1904. Chamberlin and Moulton argued that stars possess intrinsic eruptive tendencies – witness the solar flares and prominences – and that when two stars pass close to

Fig 2. James E. Keeler (Courtesy of Lick Observatory).

one another the tidal pulls between them may cause the gravitational forces restraining the eruptive tendencies to be neutralised at two points, allowing matter to be ejected. Moulton calculated that this material would be scattered along spiral arms and, because the eruption is not continuous, would be ripped out in blobs. Then, by the gradual accretion of the emitted particles whose orbits have crossed (the *planetesimals*), planetary bodies are built up.[30] Chamberlin and Moulton turned to Keeler's photographs for support for their hypothesis: no object on Keeler's plates resembled a Laplacian nebula, but they did exhibit many spiral nebulae with condensations strewn along spiral arms that seemed as if they might develop into planets. However, neither of them was of the opinion that all the spirals would form planetary systems since even if the distances to the larger and brighter spirals were relatively small, this would mean that they were far too big to be solar systems in the making. For example, in 1907 the Swedish astronomer Karl Bohlin announced that he had measured the trigonometric parallax of the Andromeda Nebula and that it indicated a distance to the spiral of about 19 light years.[31] For some time Bohlin's result was taken seriously, and since the apparent size of the Nebula was taken to be about 2°, this implied a linear size of roughly 0.7 light years, an uncomfortably, if not impossibly, large value for a bright gas cloud condensing into a system of planets and surrounding a single star.[32]

While many astronomers, particularly those outside the United States, had strong reservations about the planetesimal hypothesis, spirals were nevertheless generally seen as the first link in the chain of stellar evolution, if not as the birthplaces of single stars, then as small clusters. As astronomers had long been committed to evolutionary explanations of astronomical phenomena, it was thus unacceptable to set the spirals apart from the rest of creation and because of their cloud-like form they found their place in the cosmos as the first stage in the evolution of the stars. This situation, however, was soon to change dramatically with spiral nebulae being spoken of as possible island universes.

Spectra of spiral nebulae

In 1898 the German astronomer Julius Scheiner had secured a photograph of the spectrum, or spectrogram, of the Andromeda Nebula. Astronomers had known for many years that the spirals possessed a faint continuous spectrum, but Scheiner detected dark lines that crossed the continuous band. Further, a comparison of this spectrum with that of the Sun, taken with the same apparatus, revealed what he called a 'surprising agreement' between the two. For Scheiner the inference to be drawn was obvious: the spiral nebulae are star systems. While in the report of his observations he did not state specifically that the Andromeda Nebula is an external galaxy, the idea is clearly implied in his remark that 'the thought suggests itself of comparing the spirals with our stellar system, with especial reference to its great similarity to the Andromeda Nebula'.[33] In 1899 Huggins disclosed that after many years of examining the Andromeda Nebula, he too was convinced that its spectrum contains dark bands or groups of absorption lines. As early as 1888 Huggins had possessed spectrograms of the Nebula which had exhibited such features, but he had not made them public because he had been 'very uncertain whether the dark lines were really due to the Nebula itself, or were not rather produced by some faint traces of solar light; though the nights on which they were taken were free from moonlight at the times of taking the photographs'.[34] Despite Huggins's eminence and Scheiner's reputation as a 'careful and laborious investigator in spectroscopic astronomy'[35] and as the author of the celebrated *Die Spectralanalyse der Gestirne*,[36] their novel observations generated little excitement. Even as late as 1908, such a well-informed astrophysicist as G. E. Hale could write that 'the very numerous spiral nebulae appear to have a merely continuous spectrum'.[37] However, this belief was soon to be made untenable by E. A. Fath at Lick Observatory.

In Fath's hands the 36-inch Crossley Reflector once again proved a most formidable tool in nebular studies. With its powerful aid Fath discovered absorption lines in a number of spiral nebulae, thereby bolstering Scheiner's speculation that all spirals display them. (After Keeler's studies of nebulae between 1898 and 1900, and until the late 1920s, the nebulae that were not of a diffuse or planetary type were often classed together as 'spirals'. We shall follow this usage.) Hale was impressed enough by Fath's investigations to invite him to take a post at Mount Wilson Observatory. Fath accepted Hale's offer, and, using a 60-inch reflector and what he sadly termed a 'makeshift' spectrograph, he continued his exploration of spiral spectra until he left the Observatory four years later in 1912.[38]

Another who obtained spectrograms of spiral nebulae was the German pioneer of photographic methods in astronomy, Max Wolf. In 1912, Wolf announced that he had confirmed Fath's finding that the spirals possess an absorption line spectrum with bright lines sometimes present.[39] Bright lines were known to be the characteristic of gaseous nebulae and certain kinds of star, but their presence in the spectrum of a number of spirals was not viewed as a serious embarrassment for the island universe theory. Perhaps this was because, as it was generally accepted that gaseous nebulae are within our own Galaxy, it could be argued that it is only to be expected that stars known to exhibit emission lines and gaseous nebulae should co-exist in an external galaxy and that such stars and nebulae would sometimes reveal themselves by their emission lines in the galaxy's spectrum. Wolf, however, was not making much progress with his researches. In June 1914 he lamented that, besides the little time he had available for scientific studies, last winter 'I got never a good spectrum in spite of many trials. My spectroscope is too bad, and the same happens with our climate and my health. I fear that I have to give up all work in this direction.'[40] Wolf's investigations, as we shall see, were to be the last important observational contributions to the island universe debate by an astronomer working outside the United States.

Distances to the spiral nebulae

The 1910s saw an increasing desire on the part of astronomers to observe spiral nebulae, and to measure their distances. In 1911, the American physicist and astronomer F. W. Very advanced two methods of calculating the distance to the Andromeda Nebula. In the first, he assumed that the Nebula has the same dimensions as the Milky Way. Then comparing the apparent and supposed absolute sizes of the Nebula, was led by this tautological argument to a distance of 3800 light years. Secondly, he proposed that the 1885 nova, S Andromedae, was as intrinsically bright as a galactic nova whose distance had been previously estimated, Nova Persei. A comparison of the apparent magnitudes of Nova Persei and S Andromedae then furnished a distance of 1600 light years to the Nebula. Since he had reckoned that the Galaxy is only about 120 light years in diameter (a very small estimate even for the time), he concluded that the Andromeda Nebula must lie well outside of the Galaxy's boundaries. Of the two calculational methods Very preferred the second and this analysis persuaded him that the 'white nebulae' (those with a continuous spectrum) are galaxies, and that the smallest and faintest 'may represent galaxies at a distance of one-million light years'.[41]

It is important to note here that the Galaxy was the yardstick against which suspected external galaxies were measured, and so we must bear in mind the contemporary estimates of the dimensions of the Galaxy when considering the status of the island universe theory. In the years around 1910 a size for the diameter of the Galaxy that was sometimes quoted was 30 000 light years.[42] However, there was no consensus on the matter and many astronomers preferred a much lower value, a diameter of a few thousand light years.

In 1912 Wolf was spurred by Very's investigation to make his own estimate of the distance to the spirals.[43] His scheme rested on the assumption that the Galaxy itself is a spiral; given this, Wolf then likened the dark rifts he observed in the Milky Way, the *Langshölen* and *Querhölen*, to apparently similar features in the spirals. These comparisons gave him the relative distances to a few spirals, but to determine the absolute distances he needed to calibrate his distance scale. The *Langshölen* in the Orion Nebula, a Nebula which everyone now accepted was within the Galaxy, provided the solution because he was able to fix the Nebula's distance from the stars within it. Wolf could now calculate the absolute size of the dark rift it contained. Following this method, he was able to present a table of distances to the larger spirals:

Spiral	Distance (light years)
M 31	33 000
M 33	94 000
M 81	172 000
M 101	289 000
M 51	370 000

Wolf also reckoned the diameter of an average spiral to be about 1000 light years, a value that he asserted was similar to the diameter of the Milky Way. Wolf's reasoning was of course tautological since at its very root were the assumptions that the Galaxy is a spiral and that the dark rifts in the spirals are similar to those in the Galaxy, so that in effect Wolf was assuming that the spirals are external galaxies and working out the consequences of this assumption.

In 1911, Sir David Gill, a highly regarded British astronomer, had delivered a lecture to the Royal Institution in London in which he supported the island universe theory.[44] One of the arguments that attracted him was derived from the dynamics of the Galaxy. Two years earlier Gill had confided to his old friend Hale that he inclined to the view that the cluster of stars of which our Sun is a unit, is moving in a spiral orbit – which orbit is in the plane of the Milky Way:

If this be the true hypothesis, then not unlikely the knots in the spiral nebula [M 51] are clusters of stars similar to the cluster of which our Sun is a member.

And if we viewed the cluster of which our Sun is a member say from [M 51] then we should see our Milky Way as a spiral nebula, and the [solar] cluster as a knot in that spiral.[45]

Gill's opinions carried great weight and even twenty-five years after Gill's lecture A. S. Eddington could recall that about 1910 the island universe theory had been revived and that it was advocated especially by Sir David Gill and an international group of astronomers associated with him.[46] In 1911, Eddington, soon to become the leading British astronomer, had himself spoken out in favour of the theory. Addressing the British Association he declared that although it is 'highly speculative, the [island universe theory] may help us to a possible conception of our own system. We have the bun-shaped centre consisting of uniformly distributed stars. Around this and in the same plane are coiled spirals, which correspond to the star clouds of the Milky Way'.[47] Although the hypothesis that the Milky Way is built on a spiral pattern has its origins in the nineteenth century, it owed most to the Dutch astronomer Cornelious Easton. In 1900 he had published 'A new theory of the Milky Way' in which he had argued that the most recent observations and photographs had shown that the spiral structure was much commoner in nebulae than had been supposed. Might a spiral be the plan on which the Milky Way itself is designed? To strengthen this speculation he drew a map of the Galaxy as a spiral – though he was careful to insist that the map did not *'pretend to give an even approximate representation of the Milky Way'*.[48] Easton's hypothesis was often mentioned favourably by astronomers, and in the 1910s it enjoyed a mutually beneficial relationship with the island universe theory as each was employed to support the other.

Further evidence of a resurgence of interest in visible external galaxies comes from an address in 1912 on 'The spiral nebulae' by P. Puiseux, an astronomer at the Paris Observatory. He believed the absorption line spectrum of the spirals resulted from stars clustered within them, and that the great spiral nebula M 51 had been resolved into stars and groups of stars. In consequence, Puiseux was convinced that the largest spirals were comparable with the Milky Way.[49] Also in 1912, Fath explained that although no reliable distance indicators to the spirals were known, the message of their spectrum was that the great spirals are almost certainly remote island universes, composed probably of millions of stars.[50] J. S. Plaskett, later famous for his analyses of galactic rotation, agreed. In 1911, he had reported to the Royal Astronomical Society of Canada that recent spectral studies had shown some of the nebulae to possess a solar type spectrum: 'If we consider the Great Nebula in Andromeda, which is a typical example, we are forced to the conclusion ... that it must be tens of thousands of light years distant and probably forms a universe by itself.'[51]

There were, nevertheless, many sceptics. In 1912, A. C. D. Crommelin, a staff member of the Royal Greenwich Observatory in London, asked in his column in a scientific journal, 'Are the White Nebulae Galaxies?' Crommelin described how the island universe theory was being revived to explain the properties of the white nebulae. To back his report he cited Gill's lecture and Very's paper. Crommelin himself, however, was critical since he believed it was very difficult to reconcile the theory with the fact that the supposed island universes shunned the galactic plane and clustered about the poles of the Galaxy.[52]

Fig 3. The stellar system as a spiral (S indicates the Sun). (From Easton (1900).)

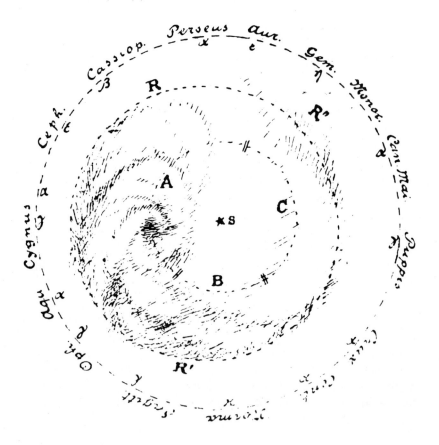

By 1913, then, external galaxies were a topical issue in astronomy, and the advocates of the island universe theory could not be dismissed easily. Very's paper had appeared in the respectable and widely read *Astronomische Nachrichten*; Puiseux's was regarded of sufficient importance for it to be translated from the *Revue Scientifique* into the *Smithsonian Report*; Eddington was just beginning to establish an international reputation; Fath was the leading researcher into the spectrum of spiral nebulae and he had observed them at two of the world's major observatories, Lick and Mount Wilson; Gill was a very eminent member of the international astronomical community; and Wolf was one of Germany's premier astronomers.

No crushing evidence was at hand to compel a shift of allegiance to the island universe theory, and to help us understand the theory's revival we shall sharpen the tools of our historical analysis and consider exactly what astronomers meant when they referred to the 'island universe theory'.[53] There were of course different views of the content of the theory, and there are two main versions that must be distinguished carefully. One stated merely that there are independent stellar systems beyond the confines of our own Galaxy. The second version might be better termed the 'comparable-galaxy theory' since this carried the extra implication that the island universes are comparable in size with the galactic system. Until the end of the nineteenth century the distinction mattered little because the great majority of astronomers wanted to force all of the unresolved nebulae into one class: *either* comparable-galaxies *or* clouds of truly nebulous matter. But in the last decades of the nineteenth century this desire for simplicity was on the wane.[54] While the shift was initially hesitant, certainly by 1910 astronomers had thrown off the shackles that had bound their nineteenth century predecessors to this restrictive belief. Also, by about 1910 it was assumed to be beyond doubt that the planetary and diffuse nebulae are within the stellar system. The evidence of the stars associated with them, their motions and their distribution in space (both the planetaries and diffuse nebulae were clearly concentrated towards the Milky Way), all pointed unambiguously towards these objects being part of the Galaxy.

By the end of the nineteenth century, as S. L. Jaki has emphasised, 'it had become customary to picture the universe as consisting of two parts: one visible and confined to the Milky Way, and another, truly infinite, which was believed to be forever beyond the reach of visible observations'.[55] Yet while astronomers had insisted that there was no observational evidence for external galaxies, some of them had been strongly inclined to think that such objects do in fact exist. In 1897 David P. Todd, Professor of Astronomy and Director of the Amherst Observatory in the United States, had speculated in his *A new astronomy* on 'Other universes than ours':

Are, then, the inconceivable vastnesses of space tenanted with other universes than the one our telescopes unfold? We are driven to conclude that in all probability they are. Just as our planetary system is everywhere surrounded by a roomy, starless void, so doubtless our huge sidereal cluster rests deep in an outer space everywhere enveloping illimitability. So remote must be these external galaxies that unextinguished light from them . . . cannot reach us in millions of years.[56]

And in 1898, J. E. Gore, a leading British writer on astronomy, had commented that the 'most probable hypothesis seems to be that all the stars, clusters and nebulae, visible in our largest telescopes, form together one vast system, which constitutes our visible universe', but he had also claimed that 'this system is isolated by a starless void from other similar systems which probably exist in infinite space'.[57] He thought this assertion could be reconciled with positive science either by invoking the extinction of light in intergalactic space, or by confining to the galaxies the luminiferous ether, that intangible and invisible substance classical physics required as a carrier of light waves, thus producing an absolute vacuum outside of the galaxies ' which would, of course, arrest the passage of all light from outer space'.[58]

Clearly, then, some, perhaps most, astronomers had been persuaded by the image of our Galaxy floating in an infinite void to fill space with as yet unseen star systems.[59] Nor was this purely a negative reaction since the island universe theory had an undoubted aesthetic appeal. For example, in 1917, A. C. D. Crommelin, now an advocate of the island universes, was to declare:

Whether true or false, the hypothesis of external galaxies is certainly a sublime and magnificent one. Instead of a single star-system it presents us with thousands of them . . . Our conclusions in Science must be based on evidence, and not on sentiment. But we may express the hope that this sublime conception may stand the test of further examination.[60]

Thus the island universe theory was an attractive theory for many, and while the observations of Fath around 1910 and the calculations of Very and Wolf were vital catalysts in the renewal of its support, a potent agent was the theory's appeal to the 'scientific taste' of its advocates. With a certain predisposition to the notion of external galaxies populating infinite space an astronomer might interpret the meagre data that had been gathered as favourable to the existence of visible island universes. Certainly in the early 1910s the observational constraints on any theory of the spirals were few, and in this situation the acceptance of different theories, each in accord with much of the available evidence, was in considerable part a matter of scientific taste. But the island universe theory's revival was nevertheless limited. With the benefit of hindsight we can see that what was required to settle the issue was a distance indicator to the spirals that was accepted as accurate by the practitioners of nebular astronomy. Until such an indicator could be fashioned, the question of the existence of island universes would remain perplexing for many.

In 1912, however, a new trail that eventually led to a novel and immensely important method of calculating the distances of spirals was opened by Vesto M. Slipher when he became the first to measure the radial velocity of a spiral nebula. Within a few years the radial velocities of numerous spirals had been secured, and in the next section we shall analyse the far-reaching implications of these measurements for the island universe theory.

Radial velocities of spiral nebulae

If a light source is in motion with respect to an observer, then the wavelengths of its spectral lines will move from the values they would have in the absence of relative motion. The amount that the lines shift, the *Doppler shift*, reveals the line-of-sight, or radial, velocity of the source. If the light source is moving towards the observer, the spectral lines are shifted towards the blue, or *blueshifted*, and if the light source is moving away from the observer the spectral lines are shifted towards the red, or *redshifted*. The first spectroscopic measurement of the radial velocity of a star was made in 1868 by William Huggins when he discovered that Sirius was fleeing from the Earth at about thirty miles a second.[61] The technique was refined by the introduction of dry photographic plates to detect the shifts of the spectral lines. These plates allowed long exposures, and so astrophysicists could now inspect far fainter stars than they could reach with the naked eye or wet plates. But even with the aid of dry plates the determination of the radial velocities of spiral nebulae proved a daunting assignment. Indeed, it was not until 1899 that any astronomer was sure even that there were lines to measure in the spectrum of a spiral. Further, the intrinsic surface brightness of the spirals is very low. Since, in contrast, most of the light of a planetary nebula was well known to be concentrated into a few bright lines, J. E. Keeler had been able in the early 1890s to measure the radial velocities of a number of planetaries.[62] The absolute values of their radial velocities were a little higher but essentially similar to those of the stars.

In 1911, W. W. Campbell, the Lick Director, derived an empirical formula linking the age of a star and its radial velocity: the older a star, the higher its speed.[63] This relationship relied on the accepted theory of stellar evolution: that, as a star ages, it changes its colour from white to red, and alters its spectral type accordingly. Since many astronomers placed the spirals at the start of the evolutionary path of the stars, they were thus expected to possess low radial velocities; in fact, of the order of 10 km s^{-1}. Even if an astronomer suspected that some of the larger spirals were not proto-stars but star clusters associated with the Galaxy, then no doubt he would have anticipated the spirals to have velocities comparable to other members of the stellar system, and certainly not far greater than those of the stars and planetary nebulae. Both of these expectations were shattered by V. M. Slipher's amazing discovery that the Andromeda Nebula is moving towards the Sun at 300 km s^{-1}, the highest speed then recorded for an astronomical body.

Slipher, a member of the Lowell Observatory at Flagstaff, had first investigated the spectrum of the Andromeda Nebula in 1909. This research programme had not been conceived by Slipher: Percival Lowell[64] kept the small staff of his observatory on a tight rein and their primary responsibility was to perform those tasks he set them.[65] Lowell's main interest was the solar system, and Mars in particular. He hoped the spirals would disclose some clues to the origin of the solar system and it was with this goal in mind that Slipher had first aimed his spectrograph at the Andromeda Nebula. By December 1910 he had detected features that, he reported to Lowell, had not been seen previously.[66] Slipher continued to try out different observing techniques, and in September 1912 his spectrograph was fitted with a very fast lens. Now he claimed that his spectrograph was about 200 times faster than the usual three prism astronomical spectrograph.[67] Within a few months he had overcome the problem of the low surface brightness of the Nebula by the use of this instrument, and by January 1913 he had four plates on which the shifts of the Nebula's spectral lines were visible. He could now measure the

Nebula's radial velocity. The combination of a skilled and determined observer, a good telescope – a 24-inch refractor – at a site with unusually clear skies, and a specially designed spectrograph had secured a remarkable result.[68]

The reaction to Slipher's find was mixed, and ranged from delight to incredulity. His astronomy teacher at Indiana University, J. A. Miller, was enthusiastic:

It looks to me as though you have found a gold mine, and that by working carefully you can make a contribution that is as significant as the one that Kepler made, but in an entirely different way.[69]

But Campbell, probably the leading American authority on radial velocity measurements, queried Slipher's result:

Your high velocity for [the] Andromeda Nebula is surprising in the extreme. I suppose, as the disperson of your instrument must have been very low, the error of your radial velocity measurement may be pretty large. I hope you have more than one result for velocity, and no doubt you have.[70]

It is not surprising that Campbell was sceptical. He was a long-standing critic of Percival Lowell, and only a few years earlier had been in the thick of a controversy over the amount of water vapour that Slipher had detected in the atmosphere of Mars.[71] At almost the same time Lowell and Campbell had also quarrelled about the relative merits of the Lowell 24-inch refractor and the Lick 36-inch refractor.[72] But any doubts about Slipher's measurement of the Andromeda Nebula's velocity faded when other astronomers confirmed the result and Slipher obtained comparable values for other spirals. In June 1914 Wolf told Slipher that his value for the Nebula's radial velocity agreed very well with Slipher's,[73] and W. H. Wright wrote to Slipher from Lick and reported that he too had corroborated the result.[74] In August 1914 Slipher delivered a paper on 'Spectrographic observations of nebulae' to a meeting of the American Astronomical Society. He had now collected the radial velocities of 15 spiral nebulae and the standing ovation given by his colleagues demonstrates the acceptance by American astronomers of his astonishing results, including two spirals that were receding at $1100 \, \text{km s}^{-1}$.[75] Three months later even Campbell told Slipher that his results on the radial velocities of the spirals 'compose one of the greatest surprises which astronomers have encountered in recent·times. The fact that there is a wide range of observed velocities – some of approach and some of recession, – lends strong support to the view that the phenomena are real. . .'[76]

The Andromeda Nebula's swift motion suggested to Slipher an explanation of the 1885 nova. He reasoned that as the Nebula raced through

Fig 4. V. M. Slipher at Lowell Observatory *c.* 1912 (Courtesy of Lowell Observatory).

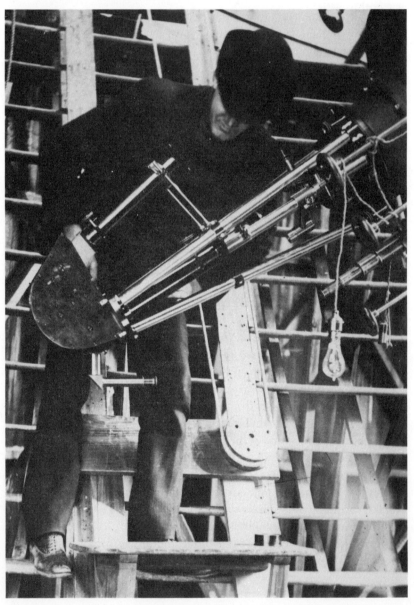

the Galaxy it may have encountered a dark star. The star's plunge into the Nebula was supposed to have sparked the processes causing the outburst that had appeared near the Nebula's nucleus.[77] This was not a new idea, but Slipher's measurements had made it more plausible. Moreover, in 1912 Slipher had proposed that the Pleiades Nebula shines by reflected starlight and he connected this hypothesis to his explanation of the 1885 nova by claiming that 'the Andromeda Nebula and similar spiral nebulae might consist of a central star enveloped and beclouded by fragmentary and disintegrated matter which shines by light supplied by the central sun. This conception is in keeping with spectrograms of the Andromeda Nebula made here and with Bohlin's value for its parallax [of 19 light years].'[78] Slipher thus explained the apparently stellar spectrum of the spirals as the product of the light of the central star or stars being reflected from the surrounding gas and dust.

The reflected-light hypothesis was soon being discussed eagerly. In June 1913 J. C. Duncan, a one time assistant at Lowell, but now Professor of Astronomy at Wellesley College, read two papers on Slipher's behalf at a meeting of the American Philosophical Society. Duncan reported to Slipher that he had spoken to E. E. Barnard and J. C. Kapteyn at the meeting and that these two leading astronomers had thought well of the hypothesis.[79] An English astronomer, J. H. Reynolds, was expounding a similar hypothesis at almost the same time as Slipher. Reynolds attempted to test the hypothesis by searching for polarisation, a characteristic of such reflected light, but he had to admit that his data were inconclusive.[80]

When Slipher delivered his 1914 address he was still not prepared to concede that the spirals are groups of stars, let alone external galaxies. Although he had been initially hesitant about interpreting the shifts of the spectral lines as Doppler shifts, he soon decided that the preponderance of redshifts really did indicate that the spirals are fleeing from the Milky Way and he suggested that this might explain why the spirals cluster around the galactic poles. He also emphasised that the spectrum of the Andromeda Nebula was without the traces of a blending that one would have expected to see in the integrated light of stars of different spectral types, and as had been found in the spectra of the globular star clusters.[81] Hence, to conclude his 1914 paper, Slipher again argued for the reflected-light hypothesis of the spirals and, in addition, he linked it to the Chamberlin–Moulton hypothesis.

A number of astronomers explained Slipher's results in a very different manner. The brilliant Danish astronomer Ejnar Hertzsprung sent to Slipher 'harty [*sic*] congratulations to your beautiful discovery of the great radial velocity of some spiral nebulae. It seems to me, that with this discovery the great question, if the spirals belong to the system of the Milky Way or not, is answered with great certainty to the end, that they do not.'[82] To Hertzsprung and others, the speeds of the spirals seemed altogether too great for them to be gravitationally bound to the stellar system, and the use of the radial velocities of the spirals to reinforce the claims of the island universe theory soon became common practice.[83] In 1916, a number of astronomers exploited the radial velocities to calculate the motion of the Sun with respect to the system of spiral nebulae. The belief underpinning these computational schemes was that the Galaxy is adrift amongst the spirals, which are themselves galaxies of stars. Just as astronomers calculated the drift of the Sun from the motions of the nearby stars, so from the motions of the spirals the drift of the Galaxy could be detected. Slipher himself performed the calculation in 1917, when he also announced his conversion to the island universe theory.[84]

Not all of the observations of the spirals, however, were seen as favouring the island universes. In 1910 G. W. Ritchey had presented a series of photographs taken with the 60-inch reflector of the Mount Wilson Observatory, then the world's largest telescope. Ritchey had a reputation as a temperamental but gifted observer, and these photographs made a deep impression on his contemporaries. His subjects had been twelve of the larger spirals and he judged that all of them contained great numbers of soft condensations. He called these 'nebulous stars', and suggested that they might be stars in the process of formation.[85] Even if the 'nebulous stars' were really stars in the making, it appeared that the spirals did not contain nearly enough for them to be galaxies comparable to the Milky Way. Ritchey's plates thereby provided ammunition that could be fired at the 'comparable-galaxy' version of the island universe theory, as well as at reflected-light hypotheses and schemes which portrayed a spiral as a single star. For example, in 1913 Duncan told Slipher:

Ritchey's recent photographs seem to show that the spirals are forming into groups of stars – rather sparse clusters instead of whole galaxies – in which case we need not be surprised that the whole nebula has the same spectrum.[86]

Globular clusters as island universes

Despite Ritchey's photographs the island universe theory continued to win supporters in the mid-1910s. One reason for this was that for a few years some astronomers identified the globular star clusters as possible island universes. Previous accounts of early twentieth century astronomy have focused upon the fortunes of the hypothesis of spiral nebulae as external galaxies, and have consequently underestimated, or missed, this aspect of the theory's revival.

Fig 5. The spiral nebula M 33 photographed by Ritchey in 1910 with the Mount Wilson 60-inch reflector. (Courtesy of Mount Wilson and Las Campanas Observatories, Carnegie Institution of Washington).

The leading student of the globular clusters *c.* 1916 was Harlow Shapley.[87] Shapley, an exceptionally gifted, ambitious and prodigiously hard working astronomer, had joined the Mount Wilson staff in 1914 after completing a doctoral thesis at the University of Princeton under the supervision of one of the United States' leading astrophysicists and the person who would continue to be his mentor, H. N. Russell.[88] Shortly after a visit to the Harvard College Observatory, Shapley had re-solved to study the stars in globular clusters while at Mount Wilson,[89] and the outcome was a classic series of papers under the general title of 'Studies based on the colors and magnitudes in clusters'. In the first paper Shapley announced the strategy of his future campaigns. He intended to tackle the clusters from an untraditional position: while con-ducting star counts, proper motion and variable star searches, as other researchers had done, he would concentrate upon the determination of the magnitudes and colours of the globular cluster stars.[90] Such measure-ments had been performed previously, but Shapley had access to the world's second largest telescope – a 60-inch reflector – and he could secure very fast exposures of these faint objects. Just as important for Shapley's globular cluster probings was the establishment of photo-graphic and photo-visual scales of magnitudes by his one time astronomy teacher but now Mount Wilson astronomer, F. H. Seares. In 1915, no scale of visual or photo-visual magnitudes existed beyond the twelfth magnitude except for the Mount Wilson sequence which Seares had ex-tended to magnitude 17.5.[91] Such scales were essential if Shapley was to extend his measures to the fainter stars in the globular clusters, and mag-nitude determinations, even to such a limit, were now relatively straight-forward but still enormously time consuming. Thus, Shapley's study of the colours and magnitudes of the faint globular cluster stars was facilita-ted by his being at Mount Wilson where the invaluable tools of the 60-inch telescope and Seares's scales were readily available to him.

Shapley was convinced that a knowledge of the distances to the 69 known globular clusters was vital to an understanding of their nature, and his persistent endeavours towards this end single him out from his contemporaries. By 1915, Shapley had already concluded that the larger the cluster, the brighter were its most luminous stars. Here was a means of deriving relative mean distances to the clusters from measures of either their diameters or the magnitudes of their brighter stars. Also, in the second paper of his series of 'Studies based on the colors and magni-tudes in clusters' he employed three techniques to estimate the distances to some globular clusters and the Small Magellanic Cloud: (1) Hertz-sprung's Cepheid variable method; (2) Kapteyn's luminosity curve; and (3) Russell's data on the absolute magnitudes of certain classes of star.[92] The Cepheid variable method relied on the relationship between the period of variation in the light emitted by a Cepheid type star and its absolute magnitude. Shapley contended that simply by observing a Cepheid's period he could deduce its absolute magnitude and hence its distance. This belief, as we shall see in Chapter 2, was vigorously disputed. The luminosity-curve method of finding distances, which had been expounded in 1914 by Kapteyn,[93] relied on a comparison of the magnitude distribution of the stars in the neighbourhood of the Sun and in the globular cluster under investigation. Shapley's third distance indicator was the absolute magnitudes of certain types of stars within the cluster. By assuming that these cluster stars have the same absolute magnitudes as stars of identical type and known absolute magnitude near the Sun, the distance to the cluster could be determined. Using these three indicators Shapley calculated the following distances:

Cluster	Distance (light years)
(1) M 13 (globular cluster)	100 000
(2) M 3 (globular cluster)	30 000
(3) ω Centauri (globular cluster)	10 000
(4) The Small Magellanic Cloud	50 000

Since Shapley believed that some of the globular clusters might be comparable to the galactic system in size (which he supposed to be probably not greater than 20 000 light years in diameter),[94] and his distances to the clusters clearly put them beyond the Galaxy, he proposed a version of the island universe theory in which the globular clusters are themselves galaxies.

That the position of the globular clusters in the cosmic hierarchy was a subject of debate elsewhere is shown by the letters of 1916 and 1917 that passed on this topic between Hertzsprung and Eddington.[95] Both men, as we have seen, were supporters of the island universe theory. Eddington had been appointed Plumian Professor of Astronomy at the University of Cambridge in 1913, and his status has been enhanced by his outstanding monograph of 1914, *Stellar Movements and the Structure of the Universe*. In this volume Eddington had again defended the island universes:

If . . . it is assumed that [the spirals] are external to the stellar system, that they are in fact systems co-equal with our own, we have at least an hypothesis which can be followed up, and may throw some light on the problems that have been before us. For this reason the 'island universe' theory is much to be preferred as a working hypothesis; and its consequences are so helpful as to suggest a distinct probability of its truth.[96]

Hertzsprung and Eddington's correspondence began because Hertzsprung wanted to hear the views of Eddington, as a leading authority on stellar dynamics, on five hypotheses that he had formed on the nature of the spiral nebulae and globular clusters. They were:

(1)   The spiral nebulae are galaxies (by which he meant they are objects comparable in size to the Milky Way).

(2)   The globular clusters are intergalactic objects, but much smaller than the Milky Way.

(3)   The number of stars is of the same order of magnitude in the globular clusters and the spirals.

(4)   The globular clusters are in a state of dynamical equilibrium, the spirals are not.

(5)   A spiral nebula is formed by two globular clusters that coalesce.[97]

The discussion that ensued centred on the fifth hypothesis, possibly because Eddington was in agreement with the other four. At a meeting of the Royal Astronomical Society in May 1916 James Hopwood Jeans, a brilliant British applied mathematician, advanced his speculations on the spiral nebulae and globular clusters.[98] Hertzsprung thought them so similar to his own views that he decided the time was ripe to put these into print, and he asked Eddington to write a digest of the main points of their correspondence for *The Observatory*, a journal of which Eddington was an editor and where Hertzsprung had no qualms about airing what he still considered were half-formed schemes.[99]

Jeans later presented his own, and by then elaborated, theory of globular clusters and spiral nebulae in his immensely influential *Problems of cosmogony and stellar dynamics*.[100] The book was published in 1919, but as it had been delayed in the press for the most part it represents his position of 1917.[101] Here Jeans argued that the spiral nebulae, which he suggested had condensed out of gigantic clouds of rotating gas, are island universes (or at least potential island universes). In line with a tendency of astronomers of the period to link a group of seemingly related objects, for example all the stars, as evolutionary variants of a basic type, Jeans envisaged the destiny of a spiral nebula as development into a star cluster. Extending this idea a little further, he continued:

It seems not unreasonable to expect that the star-clusters will be of the type we have described as 'globular' – thus we may conjecture that the observed spiral nebulae are forming star-clusters similar to observed globular star-clusters and that the observed globular clusters have originated out of spiral nebulae.[102]

Jeans then, like Hertzsprung, regarded the globular clusters as island universes intimately associated with the spiral nebulae, which are themselves island universes (we shall discuss Jeans's studies on spirals in detail later). In 1911, Fath had also tentatively connected the nebulae in an evolutionary scheme which ran from the diffuse gaseous objects, to planetary nebulae, to spiral nebulae and finally to globular clusters.[103] However, he had not buttressed and articulated his scheme with a detailed mathematical analysis as Jeans was to do, and so it had not attracted attention.

To sum up: by 1916 the hypothesis that the globular clusters are galaxies was being debated seriously, and it was a debate that fostered interest in the theory of the spirals as island universes.

The Lick school

By the mid-1910s the island universe theory had gained numerous adherents, but it was especially strongly supported at Lick Observatory. Here a school developed whose members shared, in particular, a commitment to the island universe theory, and, in general, a conservative attitude towards new developments in astronomy. W. W. Campbell, the Lick Director, was the School's most eminent member, and in December 1916 he presented the case for the island universes to the American Association for the Advancement of Science:

We are not certain how far away [the spirals] *are: we are not certain what they are.* However, the hypothesis that they are enormously distant bodies, that they are independent systems in different degrees of development, is the one which seems to be in best harmony with known facts.[104]

In the course of his address Campbell cited seven pieces of evidence for the island universe theory:

(1) The Galaxy may be a spiral: if viewed from a great distance our own stellar system might be seen as a spiral nebula. This hypothesis had been seriously proposed and was 'receiving favourable consideration'.[105]

(2) Edge-on spirals possess *dark lanes*. If the Galaxy is itself a spiral, then it is reasonable to presume that it too possesses a dark lane of obscuring matter around it in the form of a ring that blocks from view those spiral nebulae close to the galactic plane and that this lane explains

the apparent clustering of the spirals around the galactic poles.

(3)   Since there is no class of objects known to exist within the stellar system that have velocities approaching the velocities of the spirals, how can they be galactic objects?

(4)   The rotational velocities of a few spirals have been measured by Slipher and Pease. Their observations imply that the masses of the spirals are 'stupendous' and capable of producing 'probably hundreds of thousands, and possibly millions, of stars comparable in mass with our own sun'.[106]

(5)   From studies of the proper motions of the spiral nebulae there is little to indicate that they are a part of the Galaxy.

(6)   The spectra of only a few spirals have been examined, but they have the characteristics one would expect if the spirals consist chiefly of multitudes of stars.

(7)   Some astronomers believe the Magellanic Clouds to have a spiral structure, and since the Clouds are resolved easily into stars, they may be the nearest island universes.[107] Although other spirals may be island universes, they cannot be resolved because they are too distant.

In 1917, after an analysis of the distribution of the spirals on the sky and a fruitless search for spiral nebulae in the Galaxy, the Lick astronomer R. F. Sanford concluded that the spirals are outside our own system, and that they can have no intimate connection with it dynamically.[108] Also in 1917 H. D. Curtis, another Lick astronomer, presented the first published synthesis of his ideas on the spiral nebulae.[109] He acknowledged that some still regarded a spiral nebula as the forerunner of a solar system, and most of his paper was devoted to attacking this view, his weapons being the high space velocities and the peculiar space distribution of the spirals. Curtis maintained that the diffuse nebulosities are likelier candidates for the birthplaces of the stars since young stars can be seen in them, and the space velocities of the diffuse nebulosities closely match those of the stars. Moreover, no spirals are observed in the Milky Way. How, then, could the stars have formed out of the spirals? He then utilised almost the same evidence as Campbell had a few months earlier to defend his thesis that the spiral nebulae are island universes. However, he did admit that a point against the theory was that the Galaxy did not seem to possess, unlike nearly all of the spiral nebulae, a well-defined central nucleus.

Curtis, in fact, became the leading advocate of the island universe theory. His achievements in astronomy are more remarkable in the light of his undergraduate training at the University of Michigan in classical languages. But shortly afterwards his interest in astronomy was stirred, and in 1902 he became a permanent staff member at Lick. In 1910 Curtis was placed in charge of the Crossley Reflector, whereupon he continued Keeler's programme of nebular and cluster observations. At this time Curtis's views on the spirals were very different from those he was to adopt a few years later. In 1911, he contended that the globular cluster in Hercules (M 13) was a group of 25 000 stars at its densest part at least one light year across, and that such a great spiral nebula as M 51 was of a similar size.[110] However, the tone of Curtis's report to the Lick Director on his researches for the period 1st July 1913 to 15th May 1914 shows that he had now changed his mind. Here he asked rhetorically whether the nebulae are 'a part of our own stellar universe, or whether, as some astronomers have proposed in the case of the spiral nebulae, they are inconceivably distant, galaxies of stars or separate stellar universes so remote that an entire galaxy becomes but an unresolved haze of light'.[111] His observations of dark lanes in a large number of the spirals that were seen edge-on probably played a decisive part in Curtis's switch to the island universe theory. The existence of such features had been known and discussed for many years; for example, in the middle of the nineteenth century Lord Rosse and his fellow Parsonstown observers had noticed numerous dark lanes in clusters and nebulae,[112] and by the mid-1910s the term was in current usage.[113] From their presence in his photographs of edge-on spirals Curtis argued that the strange observed distribution of the spirals was a mere optical effect, a consequence of our position in the Galaxy. For if the Galaxy is a spiral, it is likely that it too will possess such a lane of material. Now, if the spiral nebulae are island universes and the Galaxy does not occupy a privileged position, then one would expect a uniform distribution of spirals over the sky; but this uniformity is destroyed by the dark lane, a ring of obscuring matter around the outer parts of the Galaxy, which blocks the light of the low lying spirals. As we have seen, Campbell had employed this hypothesis in 1916, and in 1917 it was to be the central pillar in Sanford's defence of the island universes.

Curtis had another motive for his support of the island universe theory. Woven into the avowedly empirical threads of his thought was a reverence for the 'grandeur and majesty' of the theory, and for him this may have been as important as the available observational evidence. His most explicit expression of this response came in 1924:

... there is a grandeur and majesty in the concept [of island universes] and an agreement with the general cosmical continuity expected on philosophical grounds, which is both inspiring and alluring. Few greater concepts have ever been formed in the mind of thinking man than this one, namely, – that we, the microbic inhabitants of a minor satellite of one of the millions of suns which form our galaxy, may look out beyond its confines and behold other similar galaxies, tens of thousands of light-years in diameter, each composed, like ours, of a thousand million or more suns, and that, in so doing, we are penetrating the greater cosmos to distances of from half a million to a hundred million light years.[114]

Proper and internal motions of spirals

In the nineteenth century many had tried to detect the proper and internal motions of nebulae, but it was generally believed at the end of the century that this enterprise had failed. Even in the late nineteenth century nebular photography was still in its infancy and most of these attempts had been made by comparing drawings. By the 1910s, however, astronomers believed a series of photographs of a nebula could provide an objective guide to its proper and internal motions, and the only dispute was how long it would be before such motions revealed themselves. This stimulated a number of photographic investigations of the motions of spiral nebulae, including one by Curtis. He had not been content merely with taking photographs of spirals with the Crossley telescope; he had also compared some of his plates with Keeler's taken over a decade earlier. He wanted to check: (*a*) if in the years separating the two groups of exposures there had been any perceptible motion of rotation in the spiral nebulae; and (*b*) whether the spirals show any appreciable proper motion. In 1915 Curtis completed his scrutiny. Despite an average interval of thirteen years between the two sets of plates he had not detected any evidence of internal motions, rotatory or otherwise, in the nebulae measured. However, he estimated that the average yearly proper motion of the 66 large spirals he had examined was $0''.033$.[115] Then by assuming that the spirals move randomly in space and using the method of statistical parallax, Curtis could employ Slipher's value for the mean radial velocity of the spirals to derive an average distance to them.[116] His answer was 10 000 light years, a distance which seemed to argue strongly against the island universe theory. Certainly this would have been a very surprising answer for Curtis had he trusted his proper motion data, but he did not. The poor quality of a number of Keeler's plates, a consequence of the bad mounting of the Crossley telescope at the time, and the errors involved in these determinations, the result of such difficulties as trying to measure the shift of the position of a diffuse knot of light, persuaded Curtis that the true motions were being masked and that the only worth his answer possessed was as a minimum distance to the spirals.

Curtis was not alone in his endeavour to detect proper and internal motions in spiral nebulae, and around 1915 a number of other, independent, searches were being made. Slipher's detection of the great radial velocities of the spirals had provided an extra reason for trying to measure their motions, and, before Slipher had accepted the island universe theory, he suggested to C. O. Lampland, another Lowell staff member, that Lampland hunt for motions with the Observatory's 40-inch reflector. In August 1914, Lampland announced to an American Astronomical Society meeting that he had found evidence of proper motion for the edge-on spiral NGC 4594.[117] From plates taken in 1913 and 1914 he had secured an accurate determination of the nebula's location on the sky, and he had then compared this position with eleven others with epochs stretching back to 1861 to calculate the proper motion. In May 1916, Lampland told a correspondent that 'I am at present occupied with the observation and measurement of nebulae for the determination of proper motion' and he anticipated that 'the effects of comparatively small motion should be brought out by careful measurement'.[118] Later that year he wrote that two of the largest spirals, M 51 and M 99, both exhibit detectable rotations and proper motions. In addition, the secondary nucleus of M 51 appeared to be rotating about the primary at 30 seconds of arc per century, an incredibly high figure.[119] But Lampland's results had very little influence on his contemporaries: probably his comparison of positions by use of modern and old plates, sometimes even old visually determined positions, did not seem rigorous.

If Lampland's analyses of nebular motions were ignored, then the same cannot be said of those of the Dutch astronomer at Mount Wilson, Adriaan van Maanen.[120] Van Maanen's observations were cast initially in a minor role, but they were for a few years to be at the very centre of the island universe debate.

Between 1908 and 1911 van Maanen had studied at Groningen, where he worked alongside the illustrious J. C. Kapteyn. Here he wrote his thesis on 'The proper motions of the 1418 stars in and near the clusters in h and χ Persei'. This seems to have set the course for his astronomical career for most of van Maanen's later researches were devoted to the determination of stellar parallaxes and proper motions. In 1912, after a period as voluntary assistant at Yerkes, van Maanen joined the Mount

Wilson staff. In December 1915 Ritchey wanted someone to measure two plates he had taken of the giant spiral nebula M 101, the first of them in 1910 and the second in 1915. Did the two plates, he wondered, reveal any motion of M 101? Probably because of van Maanen's experience and skill in measuring tiny displacements on photographic plates, Ritchey decided he was the man for the job. Van Maanen used a blink stereo-comparator to inspect the plates. He had used a similar machine at Yerkes and with its aid he could in effect superimpose two photographs and so compare the positions of a particular object on two plates of the same region of the sky taken at different times. To provide a reference system against which the motions of M 101 could be discerned, van Maanen chose a set of comparison stars that he hoped were not associated with the spiral. Van Maanen then selected 16 nebulous points that he assumed were within the spiral and tried to determine whether they had altered their positions in the five years separating the exposures of Ritchey's two plates.[121] Van Maanen found some signs of motion, and he decided to ask Curtis for the loan of any Crossley plates of the spiral. When in early 1916 two plates arrived at Mount Wilson from Lick he carried out a more thorough examination. The first tentative measures were confirmed and van Maanen concluded that M 101 exhibited directly measurable internal motions; he even deduced from them the spiral's rotation period, 85 000 years.

Van Maanen was delayed by Ritchey in publishing his findings. He had hoped that Ritchey, who after all had taken the Mount Wilson plates, would write the introductory section of the paper in which the results were to be announced. However, in April 1916 an exasperated van Maanen informed Hale of Ritchey's procrastinations and reported that Seares, the editor of the Mount Wilson publications, and Adams, the second in command at Mount Wilson, had told him not to wait any longer for a contribution from Ritchey.[122] The outcome was a brief paper sent to the National Academy of Sciences and a longer and more detailed one to the *Astrophysical Journal.*[123]

N. S. Hetherington has argued persuasively that van Maanen 'confirmed' rather than 'discovered' internal motions in spiral nebulae.[124] That the spirals did possess rapid internal motions did not seem to be in doubt. For example, in 1915 Curtis had contended:

... the spirals are undoubtedly in revolution since any other explanation of the spiral form seems impossible, and that the failure to find any evidence of rotation would then indicate that they must be of enormous actual size, and at enormous distances to us.[125]

Further, spectroscopic evidence that the spirals do rotate swiftly had already been secured by Slipher, Wolf and Pease.[126] Indeed, the issue was not whether the spirals do have rapid internal motions, but whether the spirals are near enough to the Earth (and small enough) for these motions to be perceptible. It was to this point that van Maanen's measurements drew attention.

Fig 6. Van Maanen's measures of the spiral M 101. The arrows indicate the direction and magnitude (greatly scaled up) of the mean annual proper motions. The comparison stars are enclosed in circles. (From van Maanen (1916a).)

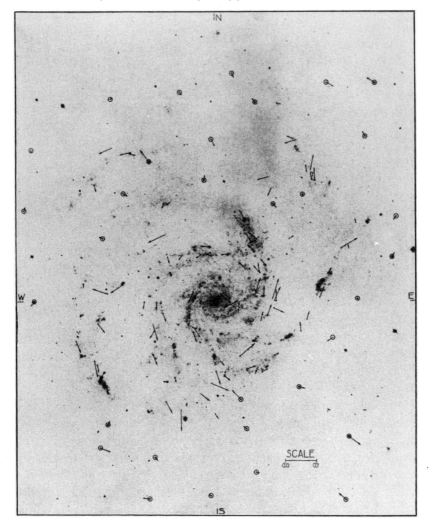

In the analysis of his data, van Maanen resolved the internal motions into radial and rotational components. He found a rotational component of $0''.022$ at an angular distance of $5'$ from the centre of the nebula. It is a simple matter to derive the formula relating these two quantities to the rotation period: $T = (2\pi/\mu_{rot})\alpha$ where α is the angular separation from the centre of the nebula and μ_{rot} is the measured proper motion caused by the nebula's rotation. Using this formula van Maanen estimated the rotation period as 85 000 years. Spectroscopic measurements of the rotational velocities of a few spirals had been made and van Maanen could now have used these values to calculate the distance of M 101 by employing the formula: $V_{rot} = (2\pi\alpha/T)d$ where d is the distance, and secured an answer of about 10 000 light years. Surprisingly, he instead quoted Curtis's characteristic distance for the spirals, which was itself 10 000 light years, but was a value whose reliability Curtis had himself seriously questioned. Van Maanen justified this move by claiming that Curtis's figure at least 'suggested' what the correct distance might be. But van Maanen did note that if he assumed an elliptical orbit for the constituents of the nebula, and applied Kepler's Third Law, then from his knowledge of the rotation period he could calculate the mass of the nebula from the formula: $M = 0.0037d^3$. In fact, if one uses this to estimate the mass of M 101, with d taken as 10 000 light years, one arrives at the staggering answer of $M \simeq 10^8$ solar masses, a figure close to contemporary values for the mass of the entire Galaxy (of order 10^9 solar masses).[127] Yet if this was anything like the actual mass of M 101, at anything like the distance van Maanen adopted, the spirals would have had a marked influence on galactic dynamics, especially on the motions of isolated stars.[128]

Maybe this remarkable result did not worry him because he preferred to view the internal motions as motions along the spiral arms rather than a rotation of the nebula. This was what one might call a common sense notion, what an astronomer may have expected to be the case on the basis of an inspection of the photographs of spirals. For example, in 1914 Eddington had said that it was clear to him that matter is either flowing into the centre of a spiral from the spiral arms or flowing out from the centre into the arms.[129] Van Maanen's measurements had to some extent justified this speculation and since the internal motions of M 101 appeared to be outwards and along the arms, probably matter was travelling to the outlying regions along them. If this were so, van Maanen reasoned that the exchange of material must affect the evolution of a spiral nebula. Van Maanen listed a group of spirals in which he noticed a gradual transition from, at one extreme, a nebula with most of its material in the centre and little in its spiral arms, to the other extreme, a nebula where most of the material resided in the arms. These were termed 'highly speculative questions' by van Maanen, but he was concerned with them 'because of the part played by elliptical motion in the Chamberlin–Moulton hypothesis as to the origin of spiral nebulae'.[130] He admitted that his findings raised difficulties for the hypothesis, although he felt that they were not insurmountable, a conclusion with which Chamberlin, a visitor to Mount Wilson in 1915, concurred.

It has been argued that when van Maanen wrote his 1916 *Astrophysical Journal* paper he had not realised the possible influence of his researches on the island universe theory.[131] This is not correct. While it is true that van Maanen was more interested (in public at least) in the significance of his results for the Chamberlin–Moulton hypothesis, he did know that they bore on the island universe theory, as a thirteen page handwritten draft of van Maanen's paper, discovered at the Hale Observatories by Hetherington, makes clear. Four paragraphs of the draft were omitted from the published paper and in one of them van Maanen wrote:

If the parallaxes of [the spiral nebulae] are really of the order of a few thousandths of a second or smaller, there is little hope that the near future will enable us to determine the distances of the spiral nebulae by the direct measuring of their parallaxes; on the other hand, the detection of their possible internal motions might indicate that these bodies are not as distant as is usually supposed . . .[132]

Why did van Maanen leave this out of his published account? Hetherington suggests that the four paragraphs were omitted because of their highly speculative nature and the scanty evidence then available;[133] but in fact three of the four paragraphs were simply revamped for the final draft, while the paragraph quoted above, which can hardly be described as 'highly speculative', was entirely suppressed. A different explanation is, however, hard to find. Perhaps the comment that 'their internal motions might indicate that these bodies are not as distant as is usually supposed' is an attack on the island universe theory; if it is, van Maanen may have had second thoughts and decided that his measurements did not clash with the existence of external galaxies as they were generally understood in 1916. Some other astronomers certainly did not think so.

In June 1916 Walter Adams, a leading astronomical spectroscopist and second-in-command at Mount Wilson, had reported to Hale:

Pease is engaged in writing up the results of his measures of the spectroscopic rotation of NGC 4594 and as soon as the paper is typewritten I will send a copy to you for your opinion. If we can assume that the angular motion derived by van Maanen and the radial motions in the nebula are of the same order, it leads to an astonishingly small parallax for the spirals – perhaps the 'other universe' hypothesis will gain support after all.[134]

Hale performed such a calculation in the Mount Wilson annual report for 1916[135] and the distance he derived, 5000 pc [parsecs], seemed to place M 101 at the boundary of, if not outside, the stellar system. Hale, nevertheless, was not convinced of this estimate's worth since he was still unsure of the reality of the measured internal motions. As Shapley disclosed to Russell in 1917, even van Maanen shared Hale's doubts to some degree: 'Do you sometimes suspect the internal proper motion in M 101? V. M. does a little, Hale a little more, and I much'.[136] Russell replied to Shapley: 'I am at present inclined to believe in the reality of the internal proper motions, and hence to doubt the island universe theory. But if [the spirals] are not star clouds, what the Dickens are they?'[137] But many others, such as Curtis, were very sceptical indeed of the motions. J. H. Jeans believed, however, that the displacements van Maanen had detected in M 101 were genuine. From about 1914 to the mid-1930s Jeans's voice was one of the most influential in astronomy and his espousal of van Maanen's motions helped them to gain acceptance. Further, Jeans was to become the leading advocate in the late 1910s and early 1920s of the harmonisation of the island universe theory with van Maanen's measures.[138]

Jeans had been led to study spiral nebulae through considering the formation of the solar system. In November 1916 he read a paper on 'The motion of tidally distorted masses' to the Royal Astronomical Society in which he stressed that the structure of the solar system could not be explained in the manner of Laplace, by a rotational theory.[139] How, then, were the planets formed? In his quest for an answer he had investigated mathematically the behaviour of bodies that exerted strong gravitational attractions on each other. Might this sort of encounter give birth to a spiral nebula? Jeans thought not because he had calculated that such interactions were very rare, and he knew there were hundreds of thousands of spirals. He thus concluded that the encounter mechanism could not apply to the vast majority of spiral nebulae, and if the spirals were not the result of such an encounter they must have their origin in some form of rotational process. In the December 1916 edition of *The Observatory* Jeans noticed a short article with the title 'Internal motions in a spiral nebula' describing van Maanen's observations of M 101. The anonymous writer (probaby Eddington) reported that the measurements were 'very accordant' and the 'motion is found to be *outwards* along the arms of the spiral. It may, alternatively, be regarded as a rotation'.[140] The rotation period of 85 000 years was also quoted, and the author stated that the spiral arms seemed to uncurl. For Jeans here was the observational vindication of his latest theoretical analyses. He quickly despatched a letter to *The Observatory*, writing:

The interesting results of Van Maanen ... are so entirely in agreement with some speculations in which I have recently been indulging, that I venture to give a brief, if somewhat premature, account of the matter.[141]

Jeans then launched into a description of his analyses of the behaviour of a rotating, shrinking mass of gas. Matter is forced out of two antipodal points by the mass's contraction and the action of external tidal forces. The ejected matter forms the arms of a spiral nebula. Moreover, the 'apparent angular velocity will clearly decrease as we pass along the arms, so that these appear to uncurl',[142] just as van Maanen had observed. In addition the nebulous knots in the arms of the spirals could be readily explained in terms of gravitationally unstable wave-motion. Then, with the aid of a few reasonable guesses and knowledge of the rotation period van Maanen had calculated for M 101, Jeans estimated the actual separation of the knots. He knew their apparent separations from their appearance on photographic plates, and by comparing these with the calculated separations, he could derive the distance to the nebula. Jeans took the Galaxy to be a lens-shaped figure of equatorial radius 2000 pc and transverse radius 600 pcs.[143] Thus, his computed distance to M 101, namely 1400 pcs, was not wholly incompatible with the island universe theory, although the low mass, 10^{37} or 5×10^{13} solar masses, told against M 101 as a comparable-galaxy.

For the next seven years Jeans was to be a champion of van Maanen's measures. Jeans had advanced his initial support in a letter to *The Observatory* after reading only a brief note about van Maanen's researches, not after a perusal of the detailed paper in the 1916 *Astrophysical Journal*. Nor does he seem to have been aware at the time of Curtis's null result for the rotation of the spirals. Even if he was, it is not surprising that Jeans was quick to back an observer whose data had seemingly embedded his own mathematical analyses in the rock of observation.

In 1917, the Adams Prize of the University of Cambridge was awarded to Jeans for an essay in which he drew together the researches he had performed on rotating, gravitating masses and the dynamics of star clusters. Two years later Jeans's essay was published as *Problems of cosmo-*

gony and stellar dynamics. The book had a mixed reception. W. D. MacMillan, a member of the Chamberlin–Moulton school at the University of Chicago, wrote a long and very critical review of it for the *Astrophysical Journal*. MacMillan protested that in the section on the solar system, Jeans had used many of Chamberlin and Moulton's ideas without giving proper credit.[144] *Problems of cosmogony and stellar dynamics* was, nevertheless, recognised immediately by many astronomers as a classic and won Jeans great prestige, particularly amongst the astronomers at Mount Wilson. For example, in 1920 Hale enthusiastically told Jeans:

We are now at work on the definitive observing programme of the 100-inch telescope, and it is hardly necessary to say that your book, which I admire so much, is our chief guide in preparing the attack on the spiral nebulae and on many other questions. I hope you will send me suggestions from time to time, as our programme will be kept elastic enough to follow them.[145]

Since *Problems of cosmogony and stellar dynamics* exerted such an enormous influence during the 1920s we shall consider it in some detail. Most of Jeans's volume was concerned with his attempt, following in the tradition of many great mathematicians, to elucidate the evolution of various rotating fluid masses. Jeans further claimed that tracing out the history of the astronomical universe is closely related to the abstract problem of following the evolution of a rotating mass of gas, and he proceeded to examine the spatial and temporal development of a gas cloud large enough to have produced a galaxy:

It has been supposed to come into existence in an entirely unknown way, probably forming at first an irregular mass of comparatively cold gas at a very low density. This will contract under its own gravitation and would in time assume a spherical shape except that it is repeatedly being disturbed by tidal forces from passing masses. The effect of these is to set up a slow rotation which continually increases as the mass contracts. The mass assumes at first a spheroidal form, then a pseudo-spheroidal form, until, when the rotation reaches [a mathematically prescribed] amount . . . a sharp edge is formed round the equator. The figure of the mass is now lenticular in shape, and any further contraction results in matter being thrown off from the periphery or equator of the lens.[146]

Because of the weak tidal forces that act on the gas cloud the matter is not shed in the form of a ring but is ejected at two antipodal points, and since material continues to flow from these two parts of the nebula, 'arms' begin to appear. As matter passes outwards, so the nebula will increase in size. By late 1916 Jeans, as we saw earlier, had already deduced relationships that enabled him to calculate the distance to, and mass of, the spiral nebula M 101. In *Problems of cosmogony and stellar dynamics* he set out similar calculations for the Andromeda Nebula. He estimated the distance to the Nebula to be 1700 pc and its mass as 10^{42} gm, or about 5×10^8 solar masses, a value comparable to those quoted for the mass of the Galaxy. Further, he calculated that the mass of a knot in the spiral arms is roughly one solar mass, an answer which bolstered his claim that they are proto-stars. Jeans warned, however, that these were merely order-of-magnitude calculations. [147]

If van Maanen's observations were correct, and matter did indeed stream out along the arms of M 101 (the so-called stream motions), then the relationship between Jeans's evolutionary scheme and the development of spiral nebulae was obvious and striking. Jeans confidently identified his abstract rotating masses with the visible spirals and he conjectured that all spiral nebulae are masses of gas or clouds of dust in rotation, this rotation being so rapid that no figure of statical equilibrium is possible. [148] But Jeans did *not* argue that all the spirals are vast collections of stars. Rather, he claimed they are aggregations of matter that are evolving into systems of stars, with stars first beginning to condense out in the arms of a spiral nebula. Jeans, like Hertzsprung, was sure that an organic connection exists between spiral nebulae and globular clusters. He maintained that his analysis led 'irresistably to the conclusion that the final result of the process of distintegration which we see going on in the spiral nebulae must be the formation of [globular] star-clusters'. [149] This conclusion 'commits us to what is commonly called the "island universe" theory'. [150] For Jeans the individual islands were the spiral nebulae and the globular clusters, but he did concede that his scheme of spirals turning into globular clusters was open to criticism. The main difficulties were twofold: there are far fewer globular clusters than spirals; and the spirals are, on average, nearer than the globular clusters. Whereas Shapley had placed the nearest globular cluster at about 20 000 light years, Curtis, Jeans wrote, had found the average distance of the spirals to be roughly 10 000 light years. Jeans had misunderstood Curtis's arguments: Curtis had proposed 10 000 light years to be the *minimum* distance to the spirals. However, Jeans was not worried about the discrepancy betwen 10 000 and 20 000 as this seemed to him a problem to be solved rather than a fatal flaw in his evolutionary hypothesis. In its defence he pointed out that the mass of the galactic system was similar to the mass he had calculated for the Andromeda Nebula and that the radial velocites of the globular clusters and the spirals were comparable.

The difference in size between the Andromeda Nebula and our Galaxy was also explained by his theory of nebular evolution: the two galaxies are not of the same dimensions because they have evolved for different lengths of time.

We have seen that van Maanen's measures of internal motions in M 101 were not unexpected, nor were they always interpreted as indicators of real motions; and those who accepted van Maanen's measures at face-value did not necessarily believe they conflicted with the island universe theory. Jeans and Adams both argued that the measures in fact *supported* the existence of external galaxies. This point must be stressed because contemporary historiography has failed to resolve the fine detail of the debate around 1917 on external galaxies, and van Maanen's measures have been presented as if they were perceived universally to be detrimental to the island universe theory. Van Maanen's measures did nevertheless destroy the belief that had been prevalent that the larger spiral nebulae are merely proto-solar-systems or sparse clusters in the making. The measures indicated masses for the spirals of at least 10^3 solar masses, and the combination of their apparent sizes and Jeans's distances (even if van Maanen's measures were exaggerations indicating only minimum distances) made the spirals appear far too large to be incipient solar systems or poorly populated clusters.

J. H. Reynolds, a tenacious opponent of the existence of external galaxies, viewed van Maanen's measures in a very different light to Jeans. Although not a professional astronomer, Reynolds possessed a 28-inch reflector sited near Birmingham, England and was for many years associated with the Helwân Observatory in Egypt where a 30-inch reflector was used to photograph nebulae between declination $-40°$ and the celestial equator. His papers were also frequently cited by the leading students of spiral nebulae.[151] In 1917, Reynolds insisted that the internal motions van Maanen had observed in M 101 definitely excluded the possibility that spiral nebulae are island universes.[152] He argued that as the size of the average proper motions found in M 101 was considerably larger than the proper motions of the faint field stars in the Galaxy, the spiral is nearer to the Sun than these stars and so inside the Galaxy. Although Reynolds realised that such a conclusion is not valid if the space velocities of the spirals are very much greater than those of the field stars, he knew that the radial velocities of a few spirals had been determined, and that they were, on average, an order of magnitude larger than those of the stars. But Reynolds thought the measured radial velocities were in error, a conclusion forced on him, in part at least, by his opposition to the island universe theory.

In 1917 Eddington wrote a short article on 'The Motions of Spiral Nebulae'. He remarked: 'The interesting question has arisen whether the spiral nebulae are comparatively small bodies scattered among the stars and belonging to the stellar system, or whether they are great subsystems on the fringe of the stellar system, or, thirdly, whether they are distant galaxies coequal with our own.'[153] If the spirals were comparable galaxies he reckoned that they must be so remote, at least 100 000 pc away, that their proper motions would have to be absolutely insensible. Curtis's observations indicated to Eddington that the existence of any detectable proper motions was very doubtful, and van Maanen's discovery of an annual proper motion of $0''.012$ for M 101 was just the size of motion that Eddington expected would accrue from the shifts of the background stars against which the motion of the spiral had been measured. However, Eddington did concede that Jeans had demonstrated that the measures fitted a possible dynamical theory.[154]

Both Eddington and Reynolds agreed that if van Maanen's measures of internal motions were genuine displacements within M 101, they would tell heavily against the island universe theory. Eddington, who accepted the existence of island universes, did not regard them as real, while Reynolds, who was hostile to the notion of external galaxies, thought them reliable.

In 1917 Reynolds and Eddington were both hampered in discussing spirals by being out of touch with the latest developments in the United States, especially at the observatories of Mount Wilson, Lick and Lowell. Neither of them, for example, was in possession of Slipher's latest listing of the radial velocities of the spirals. Nor did they realise that Slipher's measurements were trusted by American astronomers.

In the early years of the twentieth century the centre of activity in observational astrophysics had shifted in position from Europe to the United States.[155] Further, as increasing numbers of American astronomers became interested in spiral nebulae, and as the pace of the development of research programmes on spiral nebulae increased, the isolation of their colleagues outside the United States became more apparent, a situation that was exacerbated by the First World War. Scientific inquiry is guided and controlled by methods and judgements that are often informal and tacit, rather than public and explicit, and it is of just these 'informal and tacit' aspects of the scientific enterprise that an outsider has no knowledge.[156] For anyone to participate fully in the debate over island universes in the late 1910s, they had to know what was being thought and discovered in the United States, in particular at Mount Wilson, Lowell and Lick. Nor was this a matter of simply reading the relevant journals. The published papers presented only 'public science',[157] and were necessarily already somewhat out of date by the time they had been read. More important than this delay was the fact that, unless the geographical separation was to some extent compensated for by being a member of a correspondence network or through regular visits, an astronomer outside of Mount Wilson, Lick or Lowell, was likely to find himself isolated, his work ignored because it was outdated and performed with inferior instruments, his perception of theories fogged because he was ignorant of the value judgements by astronomers of other astronomers and their results, unaware too of the shifts of the criteria of theory and method selection of the leading researchers of spiral nebulae. In consequence, the observational study of the spiral nebulae had by the late 1910s become centred on the small group of astronomers tackling nebular problems at Mount Wilson, Lick and Lowell. It was a group endowed with powerful telescopes at good sites, a group that could concentrate upon such research, and it is no surprise that the next major development in the history of the island universe theory took place independently, but almost simultaneously, at Lick and Mount Wilson.

Novae in spirals

In 1917 the debate on the existence of external galaxies took another twist with the discovery of faint novae in spirals. That bright novae had appeared in two spiral nebulae was well known. The 1885 nova, S Andromedae, that had flared up in the Andromeda Nebula had been such a phenomenon. In 1895, a celestial firework on the same scale, Z Centauri, had been observed in the spiral NGC 5253, but because it was detected when rapidly fading in brightness it had failed to excite the same interest as S Andromedae.[158]

In July 1917, while taking plates of spirals for van Maanen to measure, George Ritchey detected a nova in the spiral NGC 6946.[159] Telegrams were despatched immediately to inform other astronomers of the find. 'It must have been with mixed feelings', comments M. A. Hoskin, 'that Heber D. Curtis of Lick Observatory read the telegram from Mount Wilson on 28th July', because Curtis himself was nearly ready to make a similar announcement.[160] In March 1917 he had found three possible spiral novae, one in NGC 4527 and two in NGC 4321, but he had decided to refrain from announcing any details until he was certain they were not variable stars.

After the finds of Curtis and Ritchey some astronomers were soon busily employed searching old plates for more novae. One of them was Shapley who reported to Curtis that the 'novae enthusiasm has encouraged greatly the comparison of old plates and is resulting in the discovery of a surprising number of life-like spots'.[161] Many novae were turned up, and the principal results they afforded were estimates of the magnitudes of such novae near maximum light. Curtis fastened upon this material to manufacture another distance indicator for the spirals. While he did not invent this measuring rod (S Andromedae had long been accepted as a distance indicator to M 31), the fact that a sufficient number of much fainter novae had been observed enabled Curtis to reject S Andromedae and Z Centauri on the grounds that they were anomalous objects from which conclusions should be drawn with great care. Curtis, after discussions and correspondence with Campbell, calculated the distances to the spirals in two ways. First, he argued that the invisibility of many galactic novae before and after their brief apparitions made it possible to assume that they may well have increased in brightness by at least sixteen magnitudes.[162] Then, employing the observed average magnitude of the novae in spirals, Curtis secured an average distance to the spiral novae of 20 000 000 light years by using the distance–modulus relation. In his second procedure, Curtis assumed that the spiral novae have the same absolute magnitudes as galactic novae and exploited the result that spiral novae appeared on average ten magnitudes fainter than their galactic counterparts, and so one hundred times further away. As all the available evidence indicated very large distances to the galactic novae, the spirals containing the novae must lay far outside the Galaxy. Furthermore, these spirals had relatively large angular diameters and so were 'undoubtedly' the nearer ones. For Curtis the inference to be drawn was clear: 'the novae in spirals furnish weighty evidence in favor of the well known "island universe" theory of the spiral nebulae'.[163]

In October 1917 Shapley too wrote about the spiral novae. He also deployed them in support of the island universe theory and he argued that their presence in the Andromeda Nebula indicated it to be at a distance of one million light years. Shapley in addition commented on the presence of possible gobular clusters around the Andromeda Nebula. For some years Shapley had supposed the globular clusters to be external galaxies, but in late 1917 his views on the nature of the clusters were undergoing a radical change and he was no longer prepared to see

them as independent island universes; rather, they were objects secondary to the Galaxy. With this interpretation in mind, he invited the reader
to imagine a typical globular cluster placed in a spiral that is assumed to
be 300 000 light years away: 'Then its integrated magnitude of 18 to 20,
its diffuse outline, and its apparent diameter of 3″ would simulate very
well the small nebulous condensations of spirals.'[164] Hence, the condensations around the Andromeda Nebula and other spirals might be identical to the globular clusters that were scattered around our own Galaxy.
Fifteen years later, Shapley recalled that it 'was these hazy, symmetrical
blobs that incited me to make those early comments . . . on the distances
of the nearer spiral nebulae'.[165] However, the impression given by his
1917 paper is that he regarded the novae as the strongest available evidence for the spirals as external galaxies.[166] In the light of the views
Shapley was soon to adopt, we should emphasise that Shapley, since he
now took 50 000 light years to be a minimum diameter for the Galaxy,
wrote that measurable internal motions in spirals 'can not well be har-
monised with "island universes" of whatever size, if they are composed
of normal stars';[167] that is, Shapley had decided that van Maanen's measures and the island universe theory could not be easily combined, and
when pressed he had preferred the island universes.

The novae had enabled Curtis and Shapley to construct a much desired
distance indicator to the spirals, but the scale was built necessarily from
some ill-fitting parts. First, the calculations rested on the assumption
that the novae in the Galaxy and the spirals reached similar absolute
magnitudes. Secondly, no one was sure how bright the galactic novae
were, and some astronomers, including Shapley and Russell, also asked
if there were dwarf novae in the Galaxy that would better match the
novae in spirals.[168] If there were, the distances to the spirals would have
to be reduced. Thirdly, and most ominously, the 1885 nova in the Andromeda Nebula was still very difficult to explain. But in general the novae
were seen to favour forcefully the existence of visible external galaxies.
As T. E. R. Philips remarked in a report on novae to the Annual General
Meeting of the Royal Astronomical Society in 1918, the novae give
'striking support to the theory that the spirals are "island universes".'[169]
The island universe theory, which by early 1917 had already become the
leading theory of the spirals, was pushed by the testimony of the novae
into an even stronger position. In mid-1917 the theory's active supporters included such prominent astronomers as Eddington, Jeans, Campbell, Shapley, Curtis, Slipher, Hertzsprung, Crommelin,[170] and the
brilliant Willem de Sitter.[171] There were many who wished to suspend

judgement, but the only spirited and persistent public criticism came
from J. H. Reynolds. He, however, was almost a lone figure trying to
stem a rising tide of astronomical opinion.[172]

Notes

1 Parsons (1926) 35.
2 Nichol (1848) vi. George Bond, Director of the Harvard College
   Observatory, was eager to publicise the quality of the Observatory's
   new 15-inch refractor and he claimed to have confirmed Rosse's
   'resolution' of the Orion Nebula. Bond proudly reported to Harvard's
   President that 'You will rejoice with me that the great Nebula in Orion
   has yielded to the powers of our incomparable telescope! . . . It should
   be borne in mind, that this Nebula and that of Andromeda have been
   the last strong-holds of the nebular theory; that is, the idea first
   suggested by the elder Herschel of nebulous matter in process of
   condensation into systems. The Nebula in Orion yielded not to the
   unrivalled skill of both the Herschels, armed with their excellent
   reflectors. It even defied the power of Lord Rosse's three foot mirror
   . . . I feel deeply sensible of the odiousness of comparison, but
   innumerable applications have been made to me, for evidence of the
   excellence of the instrument, and I can see no other way in which the
   public are to be made acquainted with its merits' (Bond to President
   Everett, 22 September 1847, Harvard). In the Harvard Archives there
   is also a pocket book that Bond seems to have taken to the telescope
   with him. In his entry in this volume for 21–22 September 1847 there
   are notes taken during his first examination of the Orion Nebula with
   the 15-inch refractor. Bond scrawled triumphantly: 'Resolved.
   Mottled. Abundance of Stars.' Quotations from the Bond papers are
   reproduced by permission of Harvard University Archives.
3 Hoskin (1967) 80.
4 Note in *Monthly Notices of the Royal Astronomical Society* **16,** (1856),
   139. See also Struve (1857).
5 Webb (1863) 61. The nebula is in fact associated with the original T
   Tauri variable star.
6 Hoskin (1967) 83. See also Clerke (1903) 523.
7 Huggins & Huggins (1909) 106.
8 Huggins (1865) 167. See also A. A. Cyclopaedia (1866) 91–100. Here it
   was remarked that Huggins's observations refuted the supposition 'that
   nebulae were but immensely distant systems of suns to which our own
   stellar system, with its supposed rim in the galaxy, was comparable'
   (A. A. Cyclopaedia (1866) 99).
   The two men who had done more than any others to alter the course of
   the debate on the existence of external galaxies in the middle of the
   nineteenth century were on friendly terms. R. S. Ball, one time
   assistant to Lord Rosse and later Astronomer Royal for Ireland,
   recalled in 1893 that 'I well remember going with Lord Rosse in 1866 to
   pay my first visit to Sir William Huggins at Tulse Hill' (Ball (1907) 153).
9 J. Herschel (1864).
10 Proctor (1869) 340, see also Waters (1873). By a 'zone of dispersion' he

meant a region devoid of irresolvable nebulae. In the 1850s, two of the intellectual giants of Victorian Britain, William Whewell and Herbert Spencer, had both attacked the island universe theory, one of their arguments being the clustering of the nebulae around the galatic poles. However, neither of them bolstered their criticisms with a detailed analysis of the distribution of nebulae as Proctor was to do.

11 K. Jones (1976) and Frost (1933) 45.
12 Laplace later modified his hypothesis a number of times. Its development and reception have been studied by Jaki (Jaki (1978) especially 122–45).
13 Turner (1911) 351.
14 Roberts (1888) 65.
15 Clerke (1890) 368. Miss Clerke was highly regarded by astronomers; the embryonic library of the Mount Wilson Observatory, established in 1905, consisted of two books by her and a volume of poetry.
16 Berry (1898) 405.
17 Langley (1884) 712. See also Clerke (1885).
18 Newcomb (1888) 69. For a discussion of the assumed completeness of late nineteenth century science see Badash (1972).
19 Astrophysics can be said, speaking loosely, to have been born in 1859. In that year Kirchhoff discovered that a hot gas that absorbs one special wavelength and so produces a dark line in the spectrum of a light source, will also emit the same wavelength, but adjacent wavelengths that are not absorbed will not be emitted. Laboratory analysis revealed which chemicals emitted which lines and so Kirchhoff had opened the way to the analysis of the chemical composition of distant bodies.
20 While it is almost impossible to exaggerate the importance of photography as a tool of astronomical research, its incorporation into astronomical practice was a sluggish process, and because of the faintness of the nebulae and the relative lack of interest in the nebulae in the latter half of the nineteenth century, the assimilation of photography into nebular studies was especially slow. The first photographic processes were too insensitive for serious astronomical research. The invention in 1851 of wet plates improved the situation somewhat, but even these were restricted to useful exposures of ten to fifteen minutes. However, with the introduction of the dry plate around 1875 a new era of astronomical observation dawned, an era when the light gathering ability of a telescope was almost unlimited and an era when observations of the changes in the nebulae did not have to rely on visual measures. For accounts of the development of astronomical photography see de Vaucoleurs (1961) and Davidson (1923).
21 Huggins & Huggins (1899) 16.
22 A. Hall to C. Peters, 29 April 1866, quoted by Rothenberg (1974) 230. The use of the spectroscope was also attacked by Hall. In 1870 he lamented that more observations were not being made of the satellites of Saturn, Uranus, and Neptune so that the masses of these planets could be determined to a higher accuracy. When, however, 'the novel and entertaining observations with the spectroscope have received their natural abatement and been assigned their proper place, it is to be hoped that some of the powerful telescopes recently constructed may be devoted to this class of observations [of satellites of the above

mentioned planets] where a rich and an ample field awaits the skillful observer' (Hall (1870) 372).
23 Clerke (1885) 7.
24 See Clerke (1905) Chapter 1, de Vorkin (1977*a*) and Hale (1908).
25 Keeler (1900). Although the Crossley was capable of producing very fine photographs, for several years it proved to be a temperamental instrument because of its then poor mounting.
26 Clerke (1905) 350. Keeler wrote that if 'the spiral is the form normally assumed by a contracting nebulous mass, the idea at once suggests itself that the solar system has been evolved from a spiral nebula' (Keeler (1900) 348).
27 This was principally due to its failure to account for the distribution of angular momentum within the solar system.
28 Hale (1908) 186. Hale was one of the dominant figures in astronomy in the early twentieth century: see Wright (1966).
29 He used this term in Moulton (1906).
30 Moulton (1906) 461–85.
31 Bohlin (1907). Knut Lundmark later argued that Bohlin's measurements were vitiated by hour angle errors (Lundmark (1927) 50–3).
32 Brush reports that as early as 1905 'some uncertainty about the alleged relation between spiral nebulae and planetary systems had begun to creep in' (Brush (1978) 27). In that year Moulton wrote: 'Doubtless those spirals which have been photographed are immensely larger than the one from which our system may have developed, and as a rule have relatively less massive centres' (Moulton (1905) 169). In 1907 Chamberlin also conceded that most visible spirals were too large to be proto-solar-systems (Brush (1978) 82). Brush gives a very full account of the Chamberlin–Moulton hypothesis in Brush (1978).
33 Scheiner (1899) 150. In 1899 Miss Clerke told Keeler that 'Dr. Scheiner considers all spiral nebulae to be of truly stellar composition; and I suppose their spectra are likely to differ only in detail from that of the great Andromeda ellipse. But that they are congeries of *suns*, will take a great deal more evidence than is yet forthcoming to make me believe.' (A. Clerke to J. Keeler, 8 September 1899, Lick.)
34 Huggins & Huggins (1899) 120. Huggins had been pushed into publishing this opinion by the appearance of Scheiner's paper.
35 Macpherson (1905) 234.
36 Scheiner (1890).
37 Hale (1908) 203. See also Dyson (1910) 108.
38 See Fath (1909, 1911, 1913, 1914).
39 Wolf (1912 *a*, *b*). In 1912 Fath asked Slipher of Lowell Observatory: 'Have you noticed that Wolf of Heidelberg is tackling the spectra of the spiral nebulae? He is getting some great results. It is a pleasure to see that you as well as Wolf are convinced of the value of such observations. In my opinion there is no more fruitful field of investigation when we consider the probable bearing of every scrap of information we can get on general cosmogony' (E. Fath to V. Slipher, 2 December 1912, Lowell).
40 M. Wolf to V. Slipher, 13 June 1914, Lowell.
41 Very (1911) 454. In 1912 Very decided that his estimate of the size of the Galaxy was too small by a factor of five, and so his nebular distances

needed to be increased by a factor of five. But since 'this problem lies on the utmost verge of possible solution, any answer to it must be taken with considerable latitude' (Very (1912) 376).

42 See Newcomb (1906) 53. This admittedly crude estimate was based on star-counts.

43 Wolf (1912*a*).

44 While no copy of Gill's lecture seems to have survived, A. C. D. Crommelin reported that Gill had expressed the opinion that the white nebulae are indeed galaxies (Crommelin (1912)). On Gill see Forbes (1916).

45 D. Gill to G. Hale, 13 December 1909, R.A.S.

46 Eddington (1938) 126.

47 Eddington (1912) 260. On Eddington see Douglas (1956). He was a painfully shy person with a mild and gentle presence. He was, nevertheless, tenacious in defence of views he believed to be correct.

48 Easton (1900) 157. Easton elaborated his ideas in Easton (1913).

49 Puiseux (1913) 151.

50 Fath (1912) 767.

51 J. Plaskett (1911) 265.

52 Crommelin (1912). See also Stewart (1913) 14.

53 In 1936 Edwin Hubble claimed that the term 'island universe' arose from Humboldt's use of 'Weltinseln' in *Kosmos* (Humboldt (1850) 187). In 1851 Otté translated this as 'world island' (see Humboldt (1850), Otté's translation, 201). Hubble wrote that the 'transition to "island universes" is an obvious step, but the writer has not ascertained the first use of the term' (Hubble (1936) 25). K. Glyn Jones gives the same account as Hubble of the introduction of the term 'island universe' (K. Jones (1971) 33).

54 See, for example, Miss Clerke (1903) 448. Here she supposed that the white nebulae might be called 'globular clusters in disguise'.

55 Jaki (1973) 270.

56 Todd (1897) 471. See also Comstock (1901) 355.

57 Gore (1898) 545.

58 Gore (1898) 546. The physics of the interaction at the aether–vacuum boundary was not really thought out: see Jaki (1969) 185.

59 Another problem that astronomers of the nineteenth century and early years of the twentieth, who contemplated the large scale properties of the universe, were generally aware of was the so-called Olbers's Paradox. In fact, it is not a paradox, nor was it first enunciated by Olbers. The optical paradox and its gravitational counterpart are now so well known that there is no need to explain them here (see Jaki (1969) and Hoskin (1973)). We must, nevertheless, note that the paradox exerted very little influence on the island universe debate. To astronomers in the nineteenth century and the early twentieth, there seemed to be lots of ways of avoiding the paradox; that nearly all of these were erroneous and the product of less than rigorous thinking, merely demonstrates the lack of enthusiasm in these years for cosmological matters.
Along with a general dismissal of Olbers's Paradox went an uncritical belief in the infinity of space, a space that was almost invariably assumed to be Euclidean. What, after all, did it mean to choose between a finite and an infinite universe? How can one recognize an end to the universe? It seems as if astronomers adopted the

'commonsense' notion that 'if the universe is not infinite, what lies at its boundaries?', and by so doing escaped from the contemplation of such monsters of metaphysics as a finite universe. Non-Euclidean geometries offered another escape. These saw a full flowering after 1915 with Einstein's theory of general relativity. However, in 1900 Karl Schwarzschild had tried to explain the properties of the universe in terms of a closed space (Schwarzschild (1900)), and Jaki reports that a similar effort was made in 1898 by Plassmann (Jaki (1973) 279).

60 Crommelin (1917) 376.
61 Huggins (1868).
62 Keeler (1894).
63 Campbell (1911). On the then accepted theory of stellar evolution see Smith (1977*a*).
64 For an excellent account of Percival Lowell's astronomical investigations and the early years of the Lowell Observatory see Hoyt (1976). See Hoyt (1980) for an overview of Slipher's career.
65 Hoyt comments that the staff 'functioned essentially as skilled astronomical technicians' (Hoyt (1976) 130).
66 V. Slipher to P. Lowell, 3 December 1910, Lowell.
67 Hoyt (1980) 422. Slipher's quest for the radial velocities of spiral nebulae is summarised very well in Hoyt (1980) 420–8.
68 Slipher's results were presented in Slipher (1913). In fact Fath had already detected a very large blueshift for the globular cluster NGC 7089. However, he had not believed it to be a Doppler shift; rather, he suspected that the shift was instrumental in origin (Fath (1909) 74) (see also note 74, W. Wright to V. Slipher, 19 August 1914, Lowell).
69 J. Miller to V. Slipher, June 1913, quoted by J. Hall (1970) 164.
70 W. Campbell to V. Slipher, 9 April 1913, Lowell.
71 See de Vorkin (1977*b*) and Hoyt (1976) 141–5. For details of another controversy in these years involving Lowell Observatory see Hetherington (1975*a*).
72 This argument seems to have exasperated even Slipher. While he confided to Miller that he had a distaste for controversy, Slipher had decided that because the Observatory had been so heavily criticised 'I have come to the conclusion that where we can defend ourselves . . . we shall have to do it otherwise everything we publish will be discredited' (V. Slipher to J. Miller, 18 October 1908, Lowell).
73 M. Wolf to V. Slipher, 13 June 1914, Lowell.
74 W. Wright to V. Slipher, 19 August 1914, Lowell; and V. Slipher to W. Wright, September 1914, Lowell. Wright told Slipher that his result, secured from an 18 hour exposure with a 12-inch refractor, was -304 ± 10 km s^{-1}. 'I had planned to get at this work years ago when Fath got his big displacement for NGC 7078 (which he thought was instrumental), but you seem to have beaten me to it. All our displacements are negative and until some large positive ones are found I should be inclined to doubt whether these signify motion in the line of sight, for the object as a whole at any rate' (W. Wright to V. Slipher, 19 August 1914, Lowell). Wright is confused here between NGC 7078 and NGC 7089; both are bright globular clusters, but Fath had found a large blueshift for NGC 7089, not NGC 7078. See also Hetherington (1971) on radial velocity determinations of spirals.
75 J. Hall (1970) 164.

76 W. Campbell to V. Slipher, 2 November 1914, Lowell.
77 Slipher (1913) 56.
78 Slipher (1912) 27.
79 J. Duncan to V. Slipher, 9 May 1913, Lowell.
80 Reynolds (1912).
81 V. Slipher, 'Spectrographic observations of nebulae', manuscript in the Lowell Archives.
82 E. Hertzsprung to V. Slipher, 14 March 1914, Lowell.
83 For example, a future British Astronomer Royal, H. Spencer Jones, did this in H. Jones (1915).
84 Slipher (1917) 409. Hoyt suggests that Percival Lowell may have persuaded Slipher to adopt the island universe theory (Hoyt (1976) 337).
85 Ritchey (1910) 32.
86 J. Duncan to V. Slipher, 17 February 1913, Lowell.
87 On Shapley see Gingerich (1975a), Bok (1978) and Shapley (1969). The latter source, based on Shapley's reminiscences in old age, is often unreliable, but it presents a vivid impression of his character.
88 An excellent guide to Russell's career and character is provided by the articles in de Vorkin & Philip (1977).
89 In 1917 Shapley told S. Bailey, an authority on the globular clusters: 'Very much of my work on clusters has been the direct result of my conversation with you [at the Harvard Observatory] three years ago when you suggested the advantages of the Mount Wilson instrument and weather and when you expressed the hope that I would join in the study' (H. Shapley to S. Bailey, 30 January 1917, Harvard). Nevertheless, from remarks made to Russell in 1920 it seems that Shapley had started a study of the diameters of isolated stars before 'fortunately' abandoning this investigation in favour of clusters. (H. Shapley to H. Russell, 30 September 1920, Harvard). Quotations from the Shapley papers are reproduced by permission of the Harvard University Archives.
90 The colour index of a star is the difference between its visual magnitude and its photographic magnitude; that is, the difference between its magnitude in yellow light and its magnitude in blue-violet light. Thus a blue star is brighter photographically than a red star and so has a negative colour index, while a red star is brighter visually than photographically and hence has a positive colour index. By 1916 it was widely accepted that colour was a good guide to spectral type, and in 1919 H. N. Russell was to write that while a star had to be brighter than about 10th magnitude for its spectrum to be observable in detail, it was possible to determine spectral types to roughly 16th magnitude with colours (Russell (1919) 394–8).
91 Shapley (1915a) 15.
92 Shapley (1915b).
93 Kapteyn (1914) 103–14.
94 Shapley (1915b) 86.
95 These letters are amongst the Hertzsprung papers at the University of Aarhus, Denmark.
96 Eddington (1914) 243. In 1915 Walter Adams, second-in-command at Mount Wilson, told Eddington: 'I want to express the great

appreciation in which I hold your *Stellar Movements*. It has already become much in the nature of a text-book with me and I know that many others have the same feeling' (W. Adams to A. Eddington, 11 August 1915, Hale).

97   E. Hertzsprung to A. Eddington, 6 February 1916, Aarhus.
98   RA.S. (1916*a*).
99   Eddington (1916*a*).
100 Jeans (1919).
101 Jeans told Shapley that the book had been held in the press since 1917 and its publication delayed by the War (J. Jeans to H. Shapley, 6 April 1919, Harvard).
102 Jeans (1919) 220.
103 Fath (1911) 60.
104 Campbell (1917) 534.
105 Campbell (1917) 531.
106 Campbell (1917) 532. Similar arguments to Campbell's had been presented by the Lick astronomer R. E. Wilson in 1915 (Wilson (1915)).
107 In 1913 Hertzsprung had exploited the Cepheid variable stars in the Small Magellanic Cloud to calculate its distance to be 30 000 light years, a distance that then placed it beyond the supposed boundaries of the Galaxy (see Chapter 2). Assuming the Large Magellanic Cloud, its apparent physical companion, was also at this distance, here was strong evidence for the existence of two island universes. However, we shall see in Chapter 2 that by about 1920 it had become widely accepted that the smallest possible size for the Galaxy was 30 000 light years, and although the Cepheid derived distance to the Small Cloud had been revised upwards, it now seemed that the clouds could not possibly be comparable galaxies. Rather, they were often viewed as mere appendages of the Galaxy.
108 Sanford (1917) 90. My thanks are due to Dr. C. D. Shane for pointing out this paper to me.
109 Curtis (1917*a*). On Curtis see Aitken (1943) and McMath (1944).
110 Curtis (1911) 162.
111 'Report 1 July 1913–15 May 1914', manuscript in the Lick Archives.
112 Observing NGC 1968 in April 1855 Rosse had noted that 'dark lanes are quite discernible in the finder eye-piece' of the 72-inch telescope (Parsons (1926) 176). He also claimed that 'a decided dark lane runs through [H 2297] in the direction of the major axis' (Parsons (1926) 180).
113 For example, in 1915 Slipher told Miller that 'for the great majority of the spindle-edge-on spirals there is a dark lane on their long diameter obviously due to absorbing or occulting material on the edge of the nebula' (V. Slipher to J. Miller, 10 December 1915, Lowell).
114 Curtis (1924) 8.
115 Curtis (1915*b*).
116 Curtis termed Slipher's value of 400 km s^{-1} 'truly enormous' (Curtis (1915*b*) 218).
117 Lampland (1918).
118 C. Lampland to J. Hagen, 22 May 1916, Lowell.
119 Lampland (1916).

120 On van Maanen see Berendzen & Shamieh (1973) and Seares (1946).

121 See Hart (1973) Chapter 4 for fuller details of van Maanen's procedures.

122 A. van Maanen to G. Hale, 29 April 1916, Hale. See also A. van Maanen to G. Hale, 2 May 1916, Hale.

123 Van Maanen (1916a,b).

124 Hetherington has argued for this thesis in Hetherington (1972) & (1975a); it is also the central thesis of Hetherington's unpublished manuscript 'Beyond the edge of objectivity'.

125 Curtis (1915a) 12.

126 Wolf (1914) 162 and Pease (1916a).

127 For example, in 1917 Jeans was to suggest that the mass of the galactic system was equal to that of 1.5×10^9 stars, on average 1.7× the mass of our Sun (Jeans (1919) 221).

128 Van Maanen performed this calculation in his short *Proceedings of the National Academy of Sciences* paper on M 101 (van Maanen (1916b)); his answer was $1.4 \times 10^8 \ M_\odot$, but was, he admitted, a very rough estimate. When he secured the distance by combining Curtis's proper motions and the motions of the stars (assuming that the spirals would be found eventually to have radial velocities similar to those of the stars), he obtained a value for the mass of M 101 of 30 000 $M_\odot$.

129 In 1914 Eddington had commented on the form of the spiral arms in spiral nebulae: although 'we do not understand the cause, we see that there is a widespread law compelling matter to flow in these forms'. Further, it is 'clear too that either matter is flowing into the nucleus from the spiral branches or it is flowing out from the nucleus into the branches' (Eddington (1914) 244).

130 Van Maanen (1916a) 225.

131 Berendzen & Hart (1973) 52.

132 Hetherington (1974a) 52.

133 *Ibid.*

134 W. Adams to G. Hale, 28 June 1916, Hale.

135 Hale (1916) 231, 255 (the non-appearance of this calculation in van Maanen's 1916 *Astrophysical Journal* paper is thus perplexing, especially as Pease gave it in a paper on NGC 4594: see Pease (1916b)).
     In this book I shall use both 'parsecs and 'light years' as units of distances, and I have almost always adopted the unit chosen by the astronomer under study.

136 H. Shapley to H. Russell, 3 September 1917, Harvard.

137 H. Russell to H. Shapley, 8 November 1917, Harvard.

138 On Jeans see Milne (1952), Smith (1977b) and the references therein. The Cambridge trained Jeans had been inspired to investigate the forms and stability of rotating liquid masses by the Cambridge mathematician G. H. Darwin as early as 1902. After devoting himself chiefly to molecular and atomic physics, in 1913 he began to focus increasingly on astronomy, returning in 1914 to the problem of rotating fluid masses, a problem that had attracted first-rate mathematicians for two hundred years, and which Jeans found to be fruitful for many areas of astronomy.

139 R.A.S. (1916b).

140 Eddington (1916b) 514.

141 Jeans (1917) 60.

142 Jeans (1917) 61.
143 Jeans (1919) 222.
144 MacMillan (1920).
145 G. Hale to J. Jeans, 25 May 1920, quoted in Milne (1952) 34. Adams
     also expressed to Jeans his own admiration of Jeans's volume
     (W. Adams to J. Jeans, 1 December 1919, Hale).
146 Jeans (1919) 206.
147 Jeans (1919) 217.
148 Jeans (1919) 215.
149 Jeans (1919) 220.
150 *Ibid.*
151 On Reynolds see Johnson (1950). Reynolds will figure prominently in
     Chapters 2 and 3. In 1924 Hubble told him: 'Your work has always
     appealed to me strongly and I find myself thinking along with you and
     constantly following up suggestions which arise from your papers' (E.
     Hubble to J. Reynolds, 5 April 1924, Reynolds Papers, R.A.S.).
152 Reynolds (1917) 131.
153 Eddington (1917a) 375.
154 Eddington (1917a) 377.
155 The emergence of the United States as the dominant power in
     observational astrophysics mirrors her rise to economic power. Nor
     were these two developments independent of each other since the
     prosperity of the American manufacturing industries, along with the
     burgeoning banking and transportation concerns, thrust fantastic
     fortunes into the hands of a small number of entrepreneurs and
     landowners. Some of these people donated part of their wealth to the
     arts and sciences and the money of this elite often went to endow
     astronomical and astrophysical research: the observatories of Lick,
     Yerkes and Mount Wilson are all monuments to scientists who
     managed to wrestle successfully with the wealthy and the generous, and
     much of the research conducted at Harvard College Observatory, for
     example the Draper Catalogue of stellar spectra, was funded by private
     philanthropy. Hence 'astronomy and astrophysics, two of the least
     immediately utilitarian of the sciences, became the most richly
     endowed of all [the sciences in the United States]' (Miller (1966) 203).
     The various, and often devious, ways in which the entrepreneurs and
     landowners were cajoled into endowing astronomical research is
     discussed in detail in Miller (1970) especially Chapter 5, and Miller
     (1966).
156 See Ravetz (1973) 71–108.
157 The term 'public science' is used here in Holton's sense: see Holton
     (1973), especially 'On the thematic analysis of science', 45–68.
158 Pickering (1895). In 1909 Wolf had found a variable star near the spiral
     M 101. In fact, the 'variable' has now been identified as a supernova
     that flared in M 101's outer regions. The supernova lies far from the
     centre of the spiral, and Sandage and Tammann have suggested that
     this is the reason why it failed to stir much interest at the time. Certainly
     no-one in 1909 suggested that the 'variable' was physically *within* the
     spiral (a full account is given in Sandage & Tammann (1974) 232–4 and
     plate 4).
159 Pickering (1917).
160 Hoskin (1976a) 49. Curtis announced his finds in Curtis (1917b).

161 H. Shapley to H. Curtis, 7 August 1917, Harvard.
162 Curtis (1917c) 109.
163 *Ibid*. For over a year Curtis's work on spirals was nearly brought to a halt because he was engaged on war-work almost continuously from August 1917 to November 1918.
164 Shapley (1917) 217.
165 H. Shapley to E. Hubble, 17 May 1932, Huntington Library. Quotations from the Hubble papers are reproduced by permission of the Huntington Library, San Marino, California.
166 See also Gingerich (1975a) 347.
167 Shapley (1917) 216.
168 Estimates of the absolute magnitudes of the novae were based on the parallaxes of only a handful of stars.
169 Philips (1918) 309.
170 In 1917 Crommelin had altered his former position of opposition to the island universe theory and asserted that 'most of the evidence seems to favour the extragalactic position of spirals' (Crommelin (1917) 376).
171 de Sitter (1917) 25. See also Chapter 5.
172 For example, he took Macpherson to task (Reynolds (1916)) for remarking that 'on the whole, we may conclude that, while certainty has by no means been reached, the "balance of evidence" is in favour of the view that the spirals are island universes' (Macpherson (1916) 134).

2
Shapley's model of the Galaxy and the 'Great Debate'

We have seen that by mid-1917 the island universe theory had become the most popular theory of the spiral nebulae among the practitioners of nebular astronomy. Yet the theory was soon to be severely criticised, and in this chapter we shall examine the origins and effects of the assault on the theory in the late 1910s and early 1920s. This examination is divided into two main parts. First, a brief account is given of the leading models of galactic structure in the first two decades of the century. This will include a study of Shapley's revolutionary model of the Galaxy, a model that when advanced in 1918 was seen by many as denying the existence of external galaxies of similar size to the Galaxy. Secondly, we discuss one of the most famous events in the history of twentieth century astronomy, the so-called 'Great Debate' between Shapley and Curtis in 1920. This will provide evidence of the status of the island universe theory in 1920, on the basis of which we shall contend that, despite the influence of Shapley's model of the Galaxy, it was still the dominant theory of the spiral nebulae.

Galactic models c. 1900

In 1900 astronomers were unsure of the size of the galactic system. The location of the Milky Way stars was also a puzzle, and astronomers still wondered whether or not the Milky Way is a genuine ring of stars that surrounds the Sun, which is itself located in a relatively empty central region. In addition, it was not agreed whether the Milky Way stars are as large and as bright as those near the Sun, or whether they are intrinsically less luminous.[1]

These uncertainties were expressed by the American Charles Young

55

in his influential *A text-book of general astronomy for colleges and scientific schools*. In this volume he avoided controversial statements and he was probably enunciating a general consensus when he maintained:

(1)   The large majority of the stars that can be observed are contained in a space enclosed by a round, flat disc which has a diameter roughly ten times its thickness.

(2)   The Sun is near the centre of the disc.

(3)   The distances of the remotest stars in the stellar system are uncertain, but it is certain that they must be at least some 10 000 to 20 000 light years away.

(4)   It is not clear if the Milky Way stars form a ring, inside of which is a relatively vacant area.[2]

The main authority for Young's comments was Simon Newcomb. In his essay 'The Structure of the Universe' Newcomb claimed that the Milky Way is indeed a ring of stars. Further, what 'we can say with a fair approximation to confidence is that, if we could fly out in any direction to a distance of 20 000, perhaps even 10 000 light years, we should find that we had left a large fraction of our system behind us'.[3] In the years around 1900, the ring theory of the Milky Way was one of the three principal models of the stellar system.[4] We have already met one of the others in Chapter 1: Easton's spiral model. The last member of the trio was the ellipsoid model, the product of statistical astronomy.

A principal aim of a handful of astronomers in the nineteenth century had been to deduce the actual arrangement of the stars in space from their apparent distribution on the sky, and hence the structure and size of the Galaxy. To do this successfully meant disentangling the two causes of a star's apparent brightness: its distance and its luminosity. For want of anything better, these astronomers had usually been driven to employ the assumptions of constant absolute magnitudes and constant spatial densities, both long recognised as grossly inadequate.[5] But towards the end of the nineteenth century and in the early twentieth, astronomers forged novel methodological and conceptual tools in their attempts to discern the structure of the Galaxy. Thus was born statistical astronomy. In statistical astronomy one is concerned not with individual stars but hundreds or thousands or even millions of them. As the historian–astronomer A. Pannekoek has pointed out:

[The nature of] the problems has changed with the object; we do not ask which stars, but how many stars have certain characteristics (colour, spectrum, duplicity) or certain values of the parameters (temperature, density, luminosity, magnitude). Counting supplies the measuring. The positions (in the sky or in space) do not matter, but the densities of distribution (over the sky or over space). Statistical laws of distribution are the objects and the working instruments of the astronomer who is dealing with thousands and millions of the heavenly host.[6]

The two most notable exponents of statistical astronomy in the late nineteenth century and early twentieth were J. C. Kapteyn and Hugo von Seeliger. Kapteyn's studies are by far the most important for a history of the island universe theory since they were highly regarded in the United States, especially at Mount Wilson where he was a frequent visitor and a Research Associate of the Carnegie Institution. However, the better to understand Kapteyn's analyses, we shall start by considering Seeliger's researches in the 1880s and 1890s.

The fundamental working hypothesis of the statistical astronomers was that the stellar system could be described by two functions only: the space density of stars $D = D(r, l, b)$ (r = distance, l = galactic longitude and b = galactic latitude); and a distribution function of luminosities $\phi(M)$ (M = magnitude) which is the same throughout space. Seeliger's approach consisted basically of counting the total number of stars between successive magnitude limits. If the Galaxy is filled with stars uniformly distributed and of equal luminosity, then each successive magnitude class should be almost four times as numerous as its predecessor. Instead, Seeliger found that the ratio was less than four and its actual value dependent upon galactic latitude, and he incorporated the results of his star counts within a schematic galactic model. The model was Sun centred and bounded by a surface of revolution whose axes passed through the galactic poles[7].

Seeliger's rigorous counting techniques enabled him to estimate (1) the flattening of the stellar system toward the plane of the Milky Way, (2) the relative extent of the stellar system, and (3) the general distribution of stars as a function of magnitude and galactic latitude[8]. But to secure detailed information on the actual spatial density an astronomer needed knowledge of the variations in the intrinsic brightnesses of the stars, and, in particular, the luminosity function which describes what proportion of the stars per unit volume in space have absolute magnitudes lying in successive equal intervals. So few trigonometric parallaxes of stars were known by 1900 that it was out of the question to determine the luminosity function directly.[9] Kapteyn thus developed various mathematical techniques to exploit the more easily obtained magnitude and proper motion data, notably an empirical expression for the average distance of a group of stars as a function of their magnitudes and proper motions. It was Kapteyn's great achievement that with the aid of these techniques, and given the appropriate data, he could approximate the luminosity function, and thereby calculate the dependence on distance of the space density of the stars.

Among the early products of statistical astronomy were estimates of the dimensions of the Galaxy. For example, in 1908 Kapteyn inferred that the limit of the stellar system is reached only at a distance of about 30 000 light years.[10] Yet statistical astronomers had to confront a problem which they realised might wreck their attempts to probe the Galaxy's size and structure: the determination of how much of a star's light is lost in its passage through space because of absorption. One consequence of a large amount of interstellar absorption would be to vitiate attempts to derive the density distribution from star counts, as we can see from the following table:[11]

r (pc)	r' (pc)	$D'(r')/D(r)$
100	104	0.87
500	595	0.51
1000	1410	0.26
2000	4000	0.075
4000	16 000	0.007
5000	28 000	0.002

Here $D(r)$ represents the actual density per cubic parsec at distance r; r' is the apparent distance if the absorption of 0.75 magnitudes per kpc (a value that would not have been controversial around 1910) is not taken into account, while $D'(r')$ is the corresponding apparent density distribution. The final column shows that the further one tries to peer through the fog of interstellar absorption, the more difficult it becomes to discover the correct density distribution. Hence, statistical astronomers were faced with the choice of deciding whether the observed thinning out of the stars with increasing distance was genuine, or whether it was merely the result of absorption. They were also uneasy because the observed change in density seemed to place the Sun in a nearly central, and apparently privileged, position in the Galaxy.[12]

By about 1915 many were inclined to believe that these fears had been groundless. There had been a shift towards the view that although scattered dark clouds do exist, the absorption of star light in space is insignificant. While examining the colours of the stars in the remote Hercules globular cluster Shapley had been very surprised to detect blue stars; that is, stars with negative colour indices.[13] If absorbing material had been present it would, Shapley knew, have manifested itself by reddening the light from the blue stars in the distant globular cluster, and the colour indices would have been more positive than for a sample of similar stars in the solar neighbourhood. Since this did not seem to happen, Shapley concluded that the values of the absorption constant that had recently been given by Kapteyn and others of around 1 magnitude per kpc 'must be from ten to a hundred times too large . . . and the absorption in our immediate region of the stellar system must be entirely negligible'.[14] Shapley's findings assured Kapteyn, and probably many others, that they need have no reservations about assigning a nearly central position to the Sun in the Galaxy, for, as Kapteyn announced to Hale:

One of the somewhat startling consequences of no absorption is, that we have to admit that our solar system be in or near the center of the universe, or at least to some local center.

Twenty years ago this would have made me very sceptical . . . Now it is not so – Seeliger, Schwarzschild, Eddington and myself have found that the number of stars per unit volume is greatest near the Sun. I have sometimes felt uneasy in my mind about this result, because in the derivation the consideration of scattering of light in space has been neglected. Still it appears more and more that the scattering must be too small, and also somewhat different in character from what would explain the change in apparent density. The change is therefore pretty surely real.[15]

The ironic outcome of Shapley's results on absorption was to further entrench those galactic models that placed the Sun near the centre of the Galaxy, a position from which Shapley strove to remove it three years later.

Shapley's galactic model

In 1914, in his *Stellar movements and the structure of the Universe*, Eddington described the picture that modern researches had painted of the stellar system:

It is believed that the great mass of the stars . . . are arranged in the form of a lens- or bun-shaped system. This is to say, the system is considerably flattened towards one plane . . . In this aggregation the Sun occupies a fairly central position . . . The median plane of the lens is the same as the plane marked out in the sky by the Milky Way, so that when we look in any direction along the galactic plane (as the plane of the Milky Way is called), we are looking towards the perimeter of the lens where the boundary is most remote . . . The thickness of the system, though enormous compared with ordinary units, is not immeasurably great. No definite distance can be specified, because it is unlikely that there is a sharp boundary; there is only a gradual thinning out of stars. The facts would perhaps be best expressed by saying that the surfaces of equal density resemble oblate spheroids. To give a general idea of the scale of the system, it may be stated that in directions towards the galactic poles the density continues practically uniform up to a distance of about 100 parsecs; after that the falling off becomes noticeable, so that at 300 parsecs it is only a fraction (perhaps a fifth) of the density near the Sun. The extension in the galactic plane is at least three times greater. These figures are subject to large uncertainties.[16]

Before 1918 most of the estimates of the size of the Galaxy were crudely similar to Eddington's. For example, in 1915 Shapley had remarked that the diameter of the Galaxy is probably not greater than 20 000 light years, and may be somewhat less.[17] In 1917 Curtis, following Newcomb, argued that the Galaxy is roughly lens-shaped and about 3000 by 30 000 light years in extent,[18] but many astronomers would have regarded these dimensions as over-estimates.

Imagine, then, the astonishment that greeted Shapley's announcement in 1918 that the Galaxy has a diameter of 300 000 light years, a staggering increase on the then currently accepted sizes. Moreover, Shapley placed the Sun in an eccentric position many tens of thousands of light years from the galactic centre.

The origins of this amazing theory of the Galaxy can be traced back to 1916 at least. In November of that year Shapley had confided to Eddington that while studying the galactic star cluster M 11 and the dense star clouds in its neighbourhood, he had found faint blue stars which, if they were similar to the blue stars in the solar neighbourhood, were about 50 000 light years away.[19] Further, his observations of other star clusters and star clouds had indicated analogous results to those for M 11. Shapley thus became convinced that the galactic system has a diameter of at least 50 000 light years,[20] while the sizes he had earlier derived for the globular clusters, of order 500 light years, which at the time had seemed to him to be crudely comparable to the size of the Galaxy, were far smaller than his new estimate of the diameter of the Galaxy. Shapley now decided that the globular clusters must therefore be collections of stars subordinate to the Galaxy, and not island universes in their own right. As Shapley's cluster studies had advanced he had secured what he believed to be ever more accurate distances to an increasing number of globular clusters and in October 1917 he reported to Russell that he possessed the 'distances of about thirty globular clusters – the nearest something like 20 000 light years, the furthest something like ten times as much. This is a peculiar universe.'[21] Shapley was also aware of the strange distribution of the globular clusters and in 1915 he had emphasised the well known fact that they crowd into one hemisphere of the sky.[22]

By 1917 all of the main elements for Shapley's model of the Galaxy were present: his concern for the highly asymmetrical distribution of the globular clusters across the sky; values for the distances of the clusters that he believed were reasonably accurate; a suspicion that the Galaxy was much larger than his contemporaries conceded; and a conviction that the existing galactic models were inadequate. Sometime late in 1917 these seemingly disparate elements became meshed together in Shapley's mind and he invented a remarkable galactic model: the *Big Galaxy*.

In January 1918 Shapley exclaimed to Eddington that with 'startling suddenness and definiteness' the globular clusters had elucidated the 'whole sidereal structure'.[23] Shapley reported that he had values for the distances to all the globular clusters. He had found that the equatorial

Fig 7. The globular cluster in Sagittarius M 22, photographed in 1918 with the Mount Wilson 60-inch reflector (Courtesy of Mount Wilson and Las Campanas Observatories, Carnegie Institution of Washington.)

plane of the system of globular clusters was identical with the galactic plane, and so he was now proposing that the stellar system and the huge system of globular clusters had the same centre and were co-extensive, the globular clusters actually outlining the Galaxy. Shortly after writing to Eddington, Shapley despatched a grand statement of his new view of the Galaxy to Hale as the director of his Observatory:

'(1)   The globular clusters are subordinate systems, symmetrically distributed throughout space on either side of the galactic plane.

(2)   The center of the enormous, all-comprehending galactic system is in the direction of R.A. 17 h 30 m, Decl. $-30°$ ... and its distance is of the order of 65 000 light years. The diameter of the system is some 300 000 light years in the plane. It may be much the same at right angles, but except for the first few thousand light years from the plane there is little evidence of isolated stars – only clusters, spiral nebulae, the Magellanic Clouds.

(3)   The mid-galactic region is a domain of enormous masses and tremendous gravitational forces. No globular cluster exists within 4000 light years of it, and only loose ones within 6000 light years ... Outside, besides globular clusters and nebulae, and the stars that spirals run down, we know definitely now of only two stars, but there are doubtless many more. All objects outside have exceptionally high velocities.

(4)   The globular clusters crowd up very closely to this region of avoidance on both sides. They are scattered for 200 000 light years along its boundaries. Evidently they can not form in the powerful dynamical field within those bounds where loose clusters and countless stars are; and evidently when they enter that region, or try to cross it, their fate is sealed.

(5)   But they do enter it – have entered it in the past. Within a hundred million years possibly twenty of those on my list will have tried their luck ... Do you suppose the origin of open clusters, the Hyades, Ursa Major, Messier 11, is to be found in these desperate assaults on the galactic region? ...

(6)   The dimensions of an open cluster containing only a few hundred stars is also much the same [as those of the globular clusters] but the open clusters contain only the most massive of stars.

(7)   There is no plurality of universes of which we have evidence at present (hypothesis, this is; the foregoing were conclusions). The spirals also keep out of the region of avoidance – are rushing away from it nine times out of ten ...

(8)   ... The new plan fortunately does not conflict in any way that I can see with our great store of stellar facts.

(9)   The local group, which contains all the stars we ordinarily study [everything within 1000 pc] ... is a few times larger than a globular cluster and may itself be a loose, ill-defined cluster ...
(10)   There is a maximum luminosity for stars.
(11)   The globular clusters are younger than the stars of the galactic plane; or, originating in remote regions of low forces, have developed more leisurely.'[24]

Fig 8. Shapley's 1918 comparison of distances. (From *Astrophysical Journal*, **49** (1919), 251.)

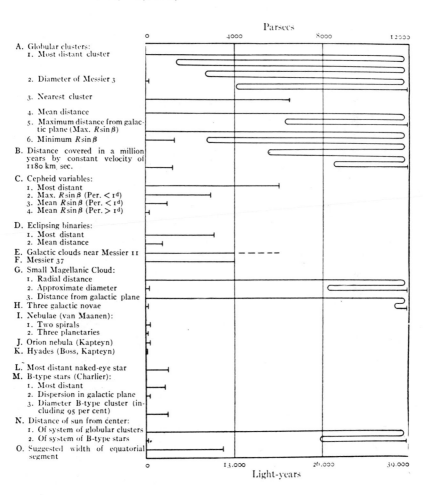

Shapley later amplified and modified the contents of this exposition. There was, nevertheless, a core to his scheme that remained essentially unaltered for a decade: that the Galaxy is a lumpy collection of star clusters; that the Galaxy is about 300 000 light years in diameter; that the Sun is eccentrically placed (these propositions in their turn depended on the various distance indicators that he had exploited); and a theory of stellar evolution that divided the stars into two basic classes, the highly luminous giants and the less luminous dwarfs. Around this hard core of the model were 'softer' layers to which Shapley was soon to make a number of changes. Set amidst these were his ideas on the spiral nebulae.

Some of the elements of Shapley's model, when taken individually, were not original. The idea that the Galaxy is constructed of a loose aggregation of star clusters, for example, was not novel. Neither did anyone suppose that the Sun was at the very centre of the stellar system, and suggestions that it was a considerable distance from the centre were not unknown. For example, in 1915 Eddington had proposed that the Sun should be placed 500 pc from the centre (he had not however expected to be believed).[25] Also, in 1909 Bohlin had proposed that the globular clusters are symmetrically placed about the centre of the galactic system and that the Sun is eccentrically positioned, a conclusion to which he had been drawn by the peculiar distribution of the globular clusters. Bohlin's paper had provoked little discussion, possibly because he argued at the same time that the planetary nebulae are external star systems,[26] a contention with which no one at the time would have agreed. But it is more likely that the lack of interest was due to the flimsy observational foundations on which his proposal rested, for, unlike Shapley, he had no firm figures for the distances to the globular clusters. In fact, in 1915 Shapley had explicitly rejected Bohlin's arrangement of the globular clusters on the grounds that it was incompatible with the distances he had derived to the globular clusters together with his 1915 estimate of the size of the Galaxy. So, in addition to advancing what were genuinely original ideas, Shapley also converted what had been vague and tenuous speculations into what he and some others regarded as bold but soundly based ideas, and placed them within the matrix of a new model of the Galaxy. True, in the construction of his model he had but one new distance indicator; but he had applied the known indicators with more determination and ruthlessness than other astronomers. These indicators had enabled him to offer empirically based solutions to specific problems, such as the distribution of the globular clusters, as well as being the base on which he erected his model of the Galaxy.

Shapley believed that the island universe theory, which he had advocated as late as October 1917, was incompatible with his new model of the Galaxy. The catalyst for this change was that he could not bring himself to regard the spirals as comparable to the huge aggregate of loosely bound clusters that he believed constituted the Galaxy. From about November 1917 onwards the vision of an immense Galaxy had grown ever larger in his thoughts until by January 1918 it dominated his thinking on external galaxies.[27] Early in 1918 he wrote:

So long as the high velocities of nebulae were unapproached by the motions of other objects and the maximum luminosity attainable by stars was beyond estimate, and so long as the diameter of the galactic system was thought to be a thousand light years or so, we had a fairly plausible case for the 'island universe' hypothesis. But now we must consider radial velocities of several hundred kilometres a second as possible for objects in our system;[28] we must assume a moderate upper limit of luminosity, perhaps even for the most massive of novae; and any external 'universe' must now be compared with a galactic system probably more than three hundred thousand light years in diameter.[29]

As well as these objections, the size of the increase in the diameter of the Galaxy that Shapley now championed threw into starker relief the problems of van Maanen's internal motions and the novae in spirals. Van Maanen's measures indicated distances of around 5000 to 10 000 light years to the larger, and presumably nearer, spirals; hence, if the Galaxy is 300 000 light years in diameter, it will easily hold these spirals within its boundaries. Shapley also contended that if the spirals were island universes, then the absolute magnitudes of some of the novae observed in them would far transcend any luminosity with which astronomers were otherwise acquainted.[30] Instead, he suggested that spiral nebulae, although not related in history or dynamical development to the average star, were members of the Galaxy.[31]

The following year, in 1919, Shapley wrote a paper 'On the existence of external galaxies',[32] containing the arguments that probably secured him his invitation to participate in the 'Great Debate'. Here Shapley explained that the internal motions in spirals detected by van Maanen, Kostinsky and Lampland[33] made diameters of 300 000 light years for the spirals impossible: the internal motions that van Maanen observed in M 101 would then indicate velocities for the outlying regions of the spiral greater than the velocity of light. For Shapley this was a blatant reduction to absurdity of the island universe theory.

Shapley further urged in his 1919 paper that only if the spirals are as near as 20 000 light years could the maximum absolute magnitude of the brightest of the spiral novae (S Andromedae) be held to a reasonable value. In addition, he suggested that the brighter spirals are within the borders of the galactic system and that the large numbers of novae detected in the bigger and more luminous spirals represent the collisions of ordinary galactic stars with the rapidly moving nebulosity that composes the spirals. The generally accepted belief that the spirals had an absorption line spectrum – a strong indication that they were composed of stars – did not trouble Shapley. The analogy with the composite spectrum of the stellar system could not, he warned, be carried very far until higher dispersion spectrographs had been trained on the spirals. Nor did the high radial velocities of the spirals present Shapley with an insurmountable difficulty. A few stars moved with radial velocities of around 400 km s^{-1}, and some globular clusters near 300 km s^{-1}, so high speed 'is not a condition impossible of production by the forces inherent in our galactic system'.[34]

Having pulled the props from under the island universe theory, Shapley was then faced with the problem of the true nature of the spiral nebulae: if they are not galaxies of stars, what are they? First, he rejected the hypothesis he had advanced in 1918 that the spirals are within the Galaxy; indeed, when in 1920 Curtis reminded him of this proposal, Shapley was astonished to find he had said such a thing.[35] In its place, he now suggested that the peculiar distribution of the spirals and their systematic recession from the Galaxy could be explained by supposing them to be masses of nebulous material repelled, in some manner, from the galactic system which he speculated to be moving through a field of such nebulae of indefinite extent.[36] Here was an imaginative vision of the Galaxy sweeping through a space full of nebulae and brushing aside those that it encountered, but assimilating the globular clusters that came within its control. However, Shapley admitted that this hypothesis was not 'competent' evidence against the island universe theory.

By 1920 many astronomers viewed Shapley's estimate of the size of the Galaxy as crucial for the fate of the island universe theory. For example, in March 1920 Shapley told Russell that Curtis, by then seen as a champion of the external galaxies, 'realizes that he must shrink my galactic system enormously to have much luck with island universes'.[37] Once again the belief that nature must be simple was influencing the island universe debate since most advocates of the island universe theory had actually adopted a comparable-galaxy theory: that is, a theory in which the external galaxies were taken to be comparable in size to the Galaxy. This assumption, which ignored the diversity in size within other classes of astronomical body, from planets to stars to clusters of stars, now created difficulties for the island universe theory and for a few years these difficulties appeared to some, as we shall see in Chapter 3, to be insuperable. Certainly if we are to comprehend the status of the theory in the late 1910s and early 1920s then we shall first have to examine the response to Shapley's model.

Shapley's model of the Galaxy was seen as a brilliant stroke by three of the most eminent and influential astronomers of the time: Eddington, Russell and Hale. By October 1918 Eddington had read the first papers in which Shapley had expounded the model, and he now declared:

I think it is not too much to say that this marks an epoch in the history of astronomy, when the boundary of our knowledge of the universe is rolled back to a hundred times its former limit.[38]

Shapley's most energetic patron was Russell, his close friend and the supervisor of his thesis studies. In May 1918 Russell wrote:

Dr Shapley was in a position to estimate the distances, in one way or another, of *all* the known globular clusters; and then to prepare diagrams showing their actual distribution in space. The results are simply amazing; but they have been derived after very careful scrutiny by an extremely competent investigator, and are thoroughly worthy of confidence.[39]

But even Shapley's most ardent supporters had some reservations. Hale had confided to Shapley that he harboured doubts about Shapley's analysis of the relationship of the spiral nebulae to the Galaxy as well as Shapley's hypothesis of the zone of avoidance (that zone of the Galaxy in which no globular clusters or spiral nebulae were visible).[40] With his second criticism, Hale had fixed on what was widely seen as a weak point in Shapley's scheme. Russell also was critical of Shapley's explanation of the zone of avoidance. He wanted proof that the absence of low lying globular clusters was not due merely to the action of obscuring matter in the galactic plane blocking such clusters from view. Nor was Russell sure that a globular cluster could evolve into an open cluster, at least in the manner claimed by Shapley. The density of stars in a globular cluster, he cautioned Shapley, is so great compared to the density in the galactic star clouds that a globular cluster would 'bore a hole through the galaxy without being greatly affected itself'.[41]

Hale too was urging Shapley to investigate the dynamical hypothesis further, although he did not want to discourage Shapley from being adventurous: 'I think you are right in making daring hypotheses, and in pushing the work ahead as you have done, as long as you . . . are prepared to substitute new hypotheses for old ones as rapidly as the evidence may demand.'[42] Hale was here being consistent with the attitude towards research he had expressed in 1915. In response to Kapteyn's profession of faith in pure induction,[43] he had replied that hypotheses should be used as guides in research; their great use, Hale insisted, is that they suggest experimental tests.[44] Hale was a spirited patron of the 'lone furrow' approach to research, and he did not treat his staff as skilled technicians whose role was to perform those tasks that he set. One should not ask an astronomer, Hale told Kapteyn in 1918, merely to perform routine cataloguing. An inductive methodology would compel such an approach, and this 'tends to turn the astronomer into a mere recording machine and discourages such splendid work as Shapley has been doing . . .'.[45] So, when in 1920 he supported Shapley's application for the vacant Directorship at the Harvard College Observatory, Hale wrote:

He certainly possesses the knowledge, ability and industry needed for the directorship of Harvard Observatory. His daring, now criticised by some, but encouraged by me in his work here, may prove to be one of his strongest qualities.[46]

There can be no denying the radical, indeed audacious, nature of Shapley's model of the Galaxy. Not surprisingly, the bold set of proposals that his model embodied met with some hostility. There are two main historical reasons for this: first, the strength of the existing amalgam of concepts, methods and theories; and second, the inherent weaknesses in Shapley's model.

Shapley's model was unexpected. There was no widespread feeling that contemporary theories of the structure and size of the Galaxy needed to be demolished and replaced by a wholly different scheme. Moreover, Shapley was arriving at his startling conclusions on the basis of a new methodology. Instead of advancing meticulously *outwards* from the region of the solar neighbourhood as did the statistical astronomers, Shapley went to the other extreme of exploiting the globular clusters to advance *into* the Galaxy.[47] Nor did the two approaches seem to link up as one would hope, and to many astronomers it seemed inherently unreasonable that the sophisticated tools of statistical astronomy could have so blunt a cutting edge that they would give such a bad value – *if* Shapley were correct – for the dimensions of the Galaxy. Shapley attempted to answer this dilemma by claiming that, although the Galaxy is composed of a collection of interacting clusters, star counts had done no more than map out the local cluster of which the Sun is a member.

W. J. A. Schouten, a practitioner of statistical astronomy and Kapteyn's last graduate student, mounted a more direct attack by criticising Shapley's distances to the globular clusters and thereby undermining Shapley's estimate of the size of the Galaxy. Schouten assumed that the distribution of stellar luminosities was identical throughout space, and in particular in the solar neighbourhood and in the globular clusters. Then, by comparing the actually observed distribution in luminosity amongst the stars in the solar neighbourhood and the globular clusters, he could obtain the distances to the globular clusters. Schouten's results indicated to him that while Shapley's model was correct in outline, the clusters were, in general, about eight times closer than Shapley believed.[48]

This point became even more pertinent when Kapteyn wrote two profound papers on galactic structure that capped his life's work in galactic astronomy. In the first, published in 1920 in collaboration with P. J. van Rhijn, Kapteyn again argued that the Sun is close to the centre of the Galaxy which the authors represented as a flattened ellipsoid.[49] Two years later Kapteyn composed a 'First attempt at a theory of the arrangement and motion of the sidereal system'.[50] Here he reckoned that the limit of the galactic system, which he took to be where the star density sank to one-hundredth of the density in the solar neighbourhood, was found roughly at 8500 pc along the galactic plane and 1700 pc at right angles to the plane. Kapteyn thereby advanced dimensions for the Galaxy that were far larger than the estimates of the mid-1910s, and so the 'Kapteyn Universe', as this model became known, was itself a notable departure from the previously prevailing orthodoxy.

Current historiography portrays a duel taking place between Kapteyn and Shapley on the canvas of galactic astronomy around 1920.[51] This black-and-white picture, however, misses the greys that were seen by their contemporaries. We shall return to this point a little later, but it is sufficient here to note that not every astronomer was prepared to back *either* Shapley *or* Kapteyn, and that the 'Kapteyn versus Shapley' description completely misses the continuing influence of the spiral theory of the Galaxy.

The second reason for the hostile reception of Shapley's model of the Galaxy by many astronomers was what they saw as weaknesses and flaws in Shapley's theorising. Because the implications of Shapley's model were so far-reaching for so many problem areas – after all he was trying to elucidate, amongst other things, the distribution of the globular clusters, the nature of the spiral nebulae and the reason for the existence of the zone of avoidance – these perceived weaknesses and flaws were many.

A study of the reception and management of novel results in science has been carried out by J. R. Ravetz. In his opinion:

...a problem under investigation grows in interaction with its materials, and that when it is completed it is necessarily rough-hewn. This will generally be even more so, in the case of deeply novel results; for the conquering of new pitfalls, the forging of new tools, and the establishment of new objects of inquiry, require great talent, daring, and ruthlessness, and also a complete identification of the scientist with his result. He will inevitably develop his own criteria of adequacy for the problem which will differ, at points likely to be crucial, from those of the established school. If every anomaly in experience, and every ambiguity in concept, were completely ironed out before the work was presented to the public, nothing new would ever appear. Also, genuinely new experimental work frequently involves using tools at, or beyond, their limits of reliability... Hence the data itself must frequently be supported by a certain measure of faith, if they are to serve as a foundation for the crucial evidence.[52]

If we here read for 'established school', 'statistical astronomy', and if for 'scientist' we read 'Shapley', then we have an excellent guide to Shapley's situation, since many astronomers believed Shapley was presenting deeply novel results by using tools – in his case distance indicators – far beyond the limits of their reliability; and probably even Shapley would have agreed that his model was 'rough-hewn'.

However, before turning to the explicit criticisms that were levelled against Shapley's model, we must examine the foundations on which he erected it – the distance indicators to the globular clusters. Shapley obtained these distances with the aid of three interlocking methods. Some of the larger (and apparently closer) globular clusters contained stars that Shapley identified as Cepheid variables. We shall discuss these stars in detail later, but all we need to know here is that Shapley argued that by observing the period of a Cepheid's light variation he could determine its intrinsic brightness and hence its distance. To reach a cluster that contained no visible Cepheids, Shapley measured the brightness of its 30 most luminous stars. He discarded the five brightest in case they were field stars and not physically associated with the cluster, and then calculated the mean luminosity of the remaining 25. Shapley could claim to know the absolute mean magnitude for these 25 stars since he had followed this same procedure for the clusters of known distance; that is, those containing visible Cepheids. By comparing the absolute and mean magnitudes he could now find the distances to clusters in which he could not detect Cepheids, but in which he could distinguish the 30 brightest stars. For the globular clusters that would not yield to either of these approaches, he employed their apparent angular diameters as distance indicators. To calibrate this final link in his chain he used the globular clusters of already determined distance to draw a curve showing apparent diameter· versus distance. These three methods enabled Shapley to secure distance estimates to all of the known globular clusters.

Shapley's first step, the Cepheid calibration, was the one to which astronomers objected vehemently and is the one we shall inspect most closely, since if he had blundered at this point, then his distances for the globular clusters would doubtless be wildly wrong, and the foundation of his galactic model destroyed.

Shapley's fundamental hypothesis was that Cepheids exhibit a relationship between the periods of their light variation and their absolute magnitudes. This had first been proposed by Miss Henrietta Leavitt at the Harvard College Observatory.[53] Miss Leavitt was a most devoted and painstaking astronomer and she spent many years scrutinising plates of the Small Magellanic Cloud that had been taken at the Harvard Southern Station. In 1908 she produced a catalogue of 1777 variables in the Cloud, commenting that while the variables fell into three or four distinct groups, most of the light curves, although of longer period, bore a striking resemblance to those of the short-period variables in the globular clusters, the 'cluster variables'. She pointed out that the brighter variables in the Cloud had the longer periods and that those with the longer periods appeared to be quite as regular in their variations as those that had periods of one or two days.[54] Sixteen variables had appeared on enough plates for Miss Leavitt to determine their periods, and four years later she possessed periods for 25. She now declared that a 'remarkable relation between the brightnesses of these variables and the length of their periods will be noticed'.[55] Further, the 'logarithm of the period increases by about 0.48 for each increase of one magnitude in brightness'.[56] Miss Leavitt realised that if the relationship could be calibrated she could use it to secure the distances to those aggregates of stars, the Small Magellanic Cloud for example, which contain these particular variables.

She did not, however, pursue this line of research,[57] and it was instead left to Ejnar Hertzsprung in 1913 to seize upon the Harvard data and exploit Miss Leavitt's discovery. The problem Hertzsprung tackled was the calibration of the relationship between period and luminosity: if the

Cepheids were to be employed as distance indicators he had to convert a
relation between periods and apparent magnitudes to a relation between
periods and absolute magnitudes. First, he identified the galactic Cepheids and the variables that exhibited the period–luminosity relation (*p–l*
relation) in the Small Magellanic Cloud as identical types of star. Next he
employed the proper motions of the 13 Cepheids with the best determined motions to calculate their mean distance. The method he used was
well-established. Since the velocities of the chosen group of stars are
assumed to be random, the average component of the proper motion in
one direction should be equal to that in any other. The Sun's motion
through space destroys this symmetry, and by observing how much it is
broken the mean distance to the group of stars can be derived. Hertzsprung thus converted Miss Leavitt's relation into one between periods
and absolute magnitudes.[58] He was now in a position to calculate the
distance to the Small Magellanic Cloud. His answer was 30 000 light
years, which in 1913 placed the Cloud beyond the supposed boundaries
of the Galaxy. Yet, since the *p–l* relation was novel and the proper
motions of the 13 Cepheids were not accurately known, this figure did
not gain wide acceptance. It transpired that in 1912 Russell had himself
determined the absolute magnitudes of the Cepheids.[59] As he later told
Hertzsprung, they had both used the same 13 stars, and had derived
practically the same absolute magnitudes for them, but Russell emphasised:

I had not thought of making the very pretty use you make of Miss Leavitt's discovery about the relation between period and absolute brightness. There is of
course a certain element of uncertainty about this, but I think it is a legitimate hypothesis. If the analogy can be pushed a little farther, one may reason as follows.
Miss Leavitt's stars, of period comparable with the Cepheid variables, are of
about the 15th photographic magnitude. The Cepheid variables appear of about
the 5th photographic magnitude. Therefore, if the two sets of stars are really
similar, and there is no absorption of light in space, the *distance of the Small
Magellanic Cloud* is 100 times of the Cepheid variables, or about 80 000 light-
years! This is an enormous distance, but is not intrinsically incredible.[60]

Hertzsprung did not, however, feel justified in classing the cluster variables as Cepheids. The cluster variables were generally known to be variable stars with light curves similar to the Cepheids; but whereas the
Cepheids had periods longer than a day, the periods of the cluster variables were less than a day. Shapley did not agree that a distinction should
be made: the subdivision of the short-period variables into the cluster

type and the Cepheid type was, he maintained, artificial. Furthermore, 'this proposition scarcely needs proof . . . Practically all writers on the subject are more or less inclined to accept this view'.[61] But Hertzsprung was not on the list of 'practically all writers', nor was Russell. In 1914 Russell commented that the cluster variables appeared to form a class independent of the longer period Cepheids, and certainly Shapley's union of the Cepheid and cluster variables was more contentious than he admitted.[62]

In 1915, in the second of the papers of his 'Studies based on the colors and magnitudes in stellar clusters' series, Shapley utilised Hertzsprung's calibration of the *p–l* curve to obtain the distances to three globular clusters and the Small Magellanic Cloud. In 1918, in the sixth paper of the series he recalibrated the *p–l* relation.[63] Although, like Hertzsprung, Shapley employed the proper motions of the Cepheids, he used a slightly different value for the solar motion and he took into account a Cepheid's change of colour index with period (this enabled him to convert photo-graphic magnitudes to visual magnitudes and so use a consistent magni-tude system). Shapley also used 11 of Hertzsprung's 13 stars, omitting κ Pavonis and l Carinae which he decided were peculiar and may be not of the Cepheid type.[64] Whereas Hertzsprung had assumed that the *p–l* slope found by Miss Leavitt for the Cepheids in the Small Magellanic Cloud was applicable to the galactic Cepheids, Shapley constructed the *p–l* slope for his 11 galactic Cepheids. To do so, he first determined the mean parallax of the Cepheids by use of the components of the proper motions parallel to the direction of the motion of the Sun through space (the ν components). Next, he calculated the mean parallax from the peculiar proper motions of the Cepheids, that is the motions that could not be accounted for simply as reflections of the solar motion. Com-bining these results, he derived the mean parallax of the 11 Cepheids and thereby the mean absolute magnitude corresponding to a particular period. He then assumed that the ν component of each star arose entirely from the Sun's motion, calculated the distance, and then the absolute magnitude, of each of the Cepheids. He could now plot the periods against the absolute magnitudes. These 11 points showed a crude re-lationship between the logarithm of the period and the absolute magni-tude, but Shapley proceeded to obtain what he termed as 'smoothed values' of absolute magnitude. First, he fitted a curve to the 11 points. From this curve he then read off the values of absolute magnitude corre-sponding to the logarithm of the period for each of the stars and drew a smoothed curve. Next, he refined the curve by adding to it the data for the Cepheids in the Small Magellanic Cloud and the globular clusters. For each collection of stars that contained Cepheids he adopted the following procedure:

(1)   From the Cepheids' periods read off the absolute magnitudes from the p–l curve.

(2)   Find the 'reduction constant'. This is the mean of the differences between the absolute magnitudes, found in (1), and the apparent magnitudes (secured by observation).

(3)   Apply the reduction constant to the apparent magnitudes to obtain new absolute magnitudes.

(4)   Plot on the p–l graph the new absolute magnitudes and periods.

(5)   Re-draw the curve to take account of these extra points.

Shapley had somehow to include in this scheme those globular clusters that contained copious cluster variables, but not longer period Cepheids. The key was the globular cluster ω Centauri since it contained both types. He derived the reduction constant for ω Centauri from the longer-period Cepheids, and used it to calculate the absolute magnitudes of the short-period cluster variables.[65] Then, because he judged the cluster variables to be generically related to longer-period Cepheids, he placed these two groups on a common p–l curve. The first section of his measuring rod to the globular clusters was thus fashioned, and Shapley could compute the distance to any globular cluster that contained visible Cepheids or cluster variables. In addition, he could now calibrate his other two distance indicators: the 30 brightest stars, and the apparent diameters of the clusters.

Yet to a number of astronomers it seemed imprudent to base a revolutionary conception of the Galaxy on a statistical analysis of the proper motions of a mere 11 stars. Further, despite the advance in accuracy achieved in Boss's *Catalogue* of proper motions, the proper motions of Shapley's 11 Cepheids were still very uncertain. Nor were Boss's values and the values of other catalogues for the proper motions of the 11 Cepheids in good agreement. Moreover, Shapley's extension of the p–l curve by linking the cluster variables to the longer-period Cepheids was a controversial move. Shapley recognised also that the trigonometrically determined distances for the Cepheids were smaller than his own distances, that the discrepancy was large and raised a question about the possible sources of error.[66] But he pushed this anomaly aside by claiming that there was evidence from an unpublished investigation by van Maanen that the trigonometric distances of the Cepheids might need to be systematically corrected, and this would bring the statistical and trigonometrically determined distances into better agreement. He further claimed that by including two other Cepheids the discrepancy would be much reduced, but this argument could not have carried much weight because he had not included the two stars in the original group. In Ravetz's terms, Shapley was prepared to back the Cepheid data with a certain measure of faith. This was not shared by many other astronomers who thought that Shapley was pushing the data beyond the limit of their reliability.

For one astronomer opposition on astronomical grounds to Shapley's galactic model was for a time reinforced by personal antagonism. Hale was the Director of the Mount Wilson Observatory until 1923, but from 1916 to 1923 he was often away from the Observatory. For some of these years he was immersed in war work, throughout the period he was

Fig 9. Shapley's 1918 period–luminosity curve for Cepheid variables. (From Shapley (1918*b*).)

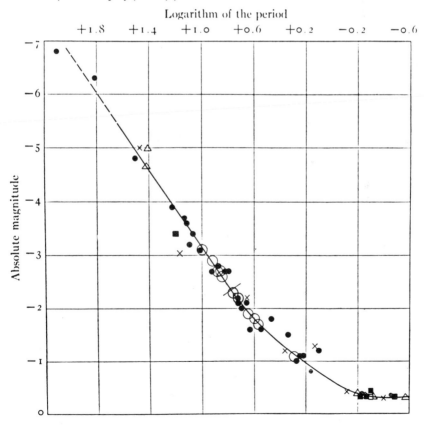

striving for the regeneration of the National Academy of Sciences, and he also suffered bouts of ill-health that kept him from the Observatory (the declining state of his health finally forced his resignation in 1923). His deputy, Walter Adams, was thus in charge of the Observatory for much of the time during these years. Shapley and Adams, however, did not see eye to eye on many matters, and Shapley was vigorous in the defence of views he believed to be correct. Charles Whitney, a Harvard trained astronomer, recalls: 'I have never seen a quicker mind, a more agile sense of humor, or a more complete absence of what usually passes for humility'.[67] These are not characteristics that would endear him to everyone, although without them it is difficult to believe that he could have advanced such an original model of the Galaxy, and then defended it so tenaciously. Shapley's supposed lack of humility seems to have annoyed the quietly efficient and conservative Adams.[68] In a letter to Hale in 1917 Adams expressed the opinion, and Hale agreed with him, that Shapley was not giving sufficient recognition to the papers and ideas of other astronomers, as well as not appraising his own studies carefully enough:

I think you have perhaps thought I have not been sufficiently appreciative of some of Shapley's recent work, particularly the magnitudes of Cepheid variables and the distances of star clusters. This has been due in part to my feeling that he has overestimated the accuracy of the results and given too little weight to other evidence, such as measured parallaxes, but more to the realization that the method is not new and that he has never given the credit where it belongs.[69]

Writing to Russell, Adams took up the theme of the discrepancy between Shapley's distances and the spectroscopically determined distances for the Cepheids.[70] Adams remarked that the spectroscopic distances of five Cepheids, three of them in the 11 used by Shapley for his 1918 Cepheid calibration, were known and their mean absolute magnitude was $+ 0.2$. Yet he said that the parallactic motions, which Shapley had used for his Cepheid calibration, were in such good agreement as to make a mean absolute magnitude of about -2.4 highly probable. Adams remarked that this difference might be explained by the small number of Cepheids Shapley had used in his calibration; nor was he sure that Shapley's statistical parallax calculations were not being confounded by non-random motions of the Cepheids.[71]

While Adams was not alone in strongly criticising Shapley's methods and results, what seems inescapable is that unlike Shapley's other critics, for a few years personal feelings invaded and coloured Adams's perception of Shapley's astronomical research. At the very least they seem to have strengthened Adam's conviction that Shapley had over-estimated the size of the Galaxy, and early in 1918 Shapley lamented to a correspondent,

I feel very sure that if I should go away from here no opportunity would be given me to return so long as Adams had the deciding voice ... van Maanen and I are in ill-favor because we do or try to do too much ... If I did not take great joy in the actual learning of things, I would feel that scientific labors are after all quite futile, for the body suffers through the necessary privations and the spirit through clashes of professional jealousies.[72]

In 1921 Shapley left Mount Wilson and joined the staff of the Harvard College Observatory; several months later he was appointed Director. During 1921 Adams's attitude towards Shapley also changed. In July he wrote to Shapley at Harvard and told him that 'I hope you will call on us freely for anything we can get you in the way of observational material. The 100-inch as you know can do some exceptional things and we ought to consider ourselves as trustees of it for the benefit of astronomers as a whole'.[73] This proffered olive-branch was greeted with a very brief reply from Shapley,[74] but in September 1921 Adams informed Shapley that he was about to write a paper and 'I wish to ask you if you have any objection to my incorporating your results on star clusters and the galaxy'.[75] Adams did not need Shapley's permission to use the cluster results; he could have cited them in the normal manner. However, Adams was offering another olive-branch to Shapley. This time Shapley grasped it; their correspondence became quite friendly and they sometimes visited each other.[76]

Nevertheless, Adams's turnabout came when Shapley had left the Observatory and possibly one of the motivations for Shapley to leave Mount Wilson was the prospect of serving under Adams. One conse-quence of his move to Harvard was that Shapley distanced himself from the centre for the studies of spiral nebulae. We shall see in Chapter 3 what effect this move had on the development of the island universe theory in the 1920s.

The 'Great Debate'

In the period between his conception of the Big Galaxy and joining the staff of the Harvard Observatory, Shapley had participated in one of the most celebrated events in the history of twentieth century astronomy: the 'Great Debate'. Before the publication in 1976 of M. A. Hoskin's 'The "Great Debate": what really happened',[77] reports of the

Debate told more of the creation of a historical romance than 'what really happened'. What is beyond dispute is that in late 1919 at a Council Meeting of the National Academy of Sciences of the United States, Hale suggested that one of the annual lectures established in memory of his father be devoted in 1920 to either island universes or relativity.[78] The outcome of this suggestion was that on 26 April 1920 Shapley and Curtis each delivered a talk to the Academy in Washington D.C. on 'The scale of the Universe'.[79] In 1921, a revised version of Curtis's talk was printed together with a paper by Shapley (which bears almost no relation to his Washington address) in a *Bulletin of the National Research Council*. These papers have been generally viewed as a '*verbatim* record of a dramatic trial of strength',[80] a notion which has been completely discredited by the discovery of the actual Debate scripts, and an analysis of the time that would have been needed to read the *Bulletin* papers.[81]

Late in life Shapley reminisced that when he heard of the death on 3 February 1919 of the Harvard Director, E. C. Pickering, he decided immediately to try and become Pickering's successor.[82] He had, however, nursed this ambition for at least a year before Pickering's death. In February 1918, at a time when Adams had prevented him borrowing some plates from Harvard, Shapley had written to an officer at Harvard:

I desire to have on record with an official concerned with the policy and future of the Harvard College Observatory a request that my name be given consideration when in the course of time the question of a successor to Director Pickering is taken up.[83]

The central thesis of Hoskin's interpretation of the Debate proceedings is that Shapley was apprehensive that his encounter with Curtis at Washington might ruin his chance of becoming the Director of the Harvard College Observatory. Shapley reasoned that if Curtis crushed his arguments he was unlikely to be offered the post, and Shapley's actions before and during the Debate, Hoskin argues, are intelligible only when we take this belief into account. (Shapley was unaware that he was not at this time being considered for the Directorship. Rather, he was being viewed as a possible right-hand-man to the Director and even when Shapley did eventually join the Harvard staff it was several months before he was to be officially appointed Director.)

Shapley was the logical choice to contest the existence of external galaxies since in 1918 and 1919 he had vigorously tried to rebut the island universe theory. Although Shapley readily agreed to be a participant in the Debate – as Hoskin points out Hale was his 'boss' and the invitation a compliment – he had strong reservations. The prospect of facing a Lick astronomer did not appeal. The Lick school contained some of the most committed opponents of his galactic model, and when Curtis seemed hesitant about being Shapley's Debate adversary, Shapley quickly suggested that an Easterner be substituted for him.[84] Moreover, Adams was not the only astronomer who had criticised Shapley for failing to cite all of his sources. In 1916, Campbell, the Lick Director, had politely laid such a charge against him because of a paper by Shapley on the spectrum of Cepheid variables.[85] Also, Shapley knew that many of the Lick astronomers did not sympathise with his daring theorising; Shapley had even invented the phrase the 'Lick state of mind'[86] to characterise what he saw as their caution.

The first person to be proposed as the advocate of the island universes was Campbell, but after Hale had seen a Lick volume on nebulae that contained three papers by Curtis it was he who became the chosen speaker. Curtis was really the spokesman for the Lick school, and before the idea of the Debate had been aired Curtis had told a correspondent that the Lick Observatory Journal Club had recently held symposia on topics including the size of the Galaxy, external galaxies and the nature of the Cepheids. 'Most of us here', Curtis reported, 'find it impossible to subscribe to some of the recent theories' on these topics[87] and several of the most recent theories were Shapley's creations. The Debate's origin even may lie in a suggestion made to Hale in March 1919 by some Lick astronomers that Shapley give a paper on his latest theories to a meeting of the National Academy of Sciences. Hale told Adams that if he liked the suggestion, 'you might ask Shapley to send a paper, or come on personally to read it if he wishes to do so'.[88]

Before the Debate Shapley was very anxious to ensure that Russell, his most influential supporter and a great force in American astronomy, would be at Washington to help fend off Curtis's thrusts.[89] A month before the Debate Shapley told Russell that Curtis intended to go after the modern views on astrophysics '"hammer and tongs" . . . with his "shillelah"'.[90] Further:

Curtis swears by Newcomb and other patriarchs, and will show (?) that my distances [to the globular clusters] are some ten times too big. Now that ten times, as Mr. Hale realizes, is as bad on your hypotheses as on mine; it is a violation of nearly all recent astrophysical theory. So unless Curtis actually bowls us over with the only true truth in these celestial matters, you will be interested in this general assault from self-styled conservatives.

Professor Brown is here at the observatory; also Professor Frost. They, as well as the people at Lick and at Mount Wilson, seem to regard the coming discussion as a crisis for the newer astrophysical theories (you know of course that Lick believes practically nothing that you or I or Eddington or Hertzsprung have done in the way of interpretation, and not much of Kapteyn's work; they play safe, and do good work in making everybody show the goods). But crisis or not, I am requested to talk to the general public of non-scientists that may happen to drop in. Consequently whatever answer must be made to Curtis and his school must be made in the discussion.[91]

While the evidence is not in harmony with Shapley's judgement of an impending crisis in astrophysics,[92] it does accord with the argument that Shapley believed the Debate would be crucial for his chances of the Harvard Directorship.

The 'Debate' took place on 26 April 1920 at the National Academy of Sciences in Washington D.C. To minimise the possibility that Curtis, an experienced public speaker, would demolish his case, Shapley, having already expended effort in undermining the seriousness of the encounter and reducing the time available for the talks, chose to speak at a non-technical level. Although members of the public were admitted to the meeting, it is hard to see what other explanation can be offered for the facts that Shapley only reached the definition of a light year after seven pages of his script of 19 pages and that he devoted the last three pages to an intensifier he had developed to photograph very faint stars. The intensifier had little bearing on the theoretical argument, but probably Shapley reasoned that it would impress those members of his audience, like Mr. Agassiz of the Harvard Observatory Visiting Committee, concerned with the future activities of Harvard College Observatory. Curtis, who had expected a technical presentation, was left throwing his verbal blows at a non-existent protagonist.[93] It was only in their correspondence after the Debate and with their papers in a *Bulletin of the National Research Council*, during the composition of which they had exchanged drafts, that Shapley and Curtis finally got to grips with each others arguments.[94]

Hale's original intention was that the Debate should centre on island universes. Before the Washington encounter, however, Shapley had confessed to Russell that 'I am going to touch lightly on the spiral nebulae (I have neither time nor data nor very good arguments)...'[95] Anyway, Shapley believed that the existence of external galaxies would stand or fall with the acceptance or rejection of his Big Galaxy. By proclaiming the correctness of his model he was by implication assailing the island universe theory, and so at Washington and in his *Bulletin* paper Shapley focused his analysis on the size of the Galaxy.

Shapley knew that Curtis reckoned his cluster distances to be about ten times too large. He decided that if he could demonstrate that one of the globular clusters was at the distance he claimed, then Curtis's argument would be refuted. What tactics should he use to achieve this strategic goal? Even before the Washington meeting he had realised that Curtis was going to attack his 11 'miserable' Cepheids,[96] and so he was wary of employing the Cepheids to fix the distance of a globular cluster.

Fig 10. H. D. Curtis *c.* 1916 (Courtesy of Lick Observatory).

In his *Bulletin* paper Shapley maintained:

...in the present issue there is little point in labouring over the details for Cepheids, for we are, if we choose, qualitatively quite independent of them in determining the scale of the galactic system, and it is only qualitative results that are now at issue. We could discard the Cepheids altogether, use instead either the red giant stars and spectroscopic methods or the hundreds of B-type stars upon which the most capable stellar astronomers have worked for years, and derive much the same distance for [Messier 13], and for other clusters, and obtain consequently similar dimensions for the galactic system.[97]

Fig 11. Harlow Shapley (Courtesy of Harvard University Archives).

His chosen tactic, then, was to outflank Curtis's assault on the Cepheids by deploying the B-type stars as his principal distance indicators.

Both Shapley and Curtis agreed that there was a uniformity of stellar conditions throughout the galactic system. In particular, Shapley claimed that the stars which he identified as B-type stars in the globular clusters were identical to the B-type stars in the solar neighbourhood and hence that they were suitable for deriving the distances to the globular clusters. Shapley tried to strengthen this identification by emphasising that in his cluster theory of the Galaxy, the near and distant B-type stars he was comparing were all cluster stars, and that there seemed to be no marked break in the gradation of clusters. Maybe Shapley preferred to use the B-type stars as his standard candles instead of the red giants because in 1918 Eddington had told Shapley that some of his cluster distances might be incorrect. Eddington had discovered contradictions between the evidence of the Cepheids and red giants:[98] he pointed out a cluster in which the red giants seemed, if the distance derived from the Cepheids was correct, to be intrinsically much brighter than the red giants in the Galaxy. Shapley had replied that he should have said that if one used the giants to determine the cluster distances, then the answers were 'of the same order' as those secured with the Cepheids and not 'closely comparable'. He confided to Eddington that he had aimed the claim of their being 'closely comparable' at the 'one or two who think the clusters are near, that the Cepheids are not giants, and that my results are insecurely founded on slim evidence'.[99]

Shapley here underestimated the opposition to his theories and his assertion that only 'one or two' people considered his results were based on meagre evidence was hopelessly optimistic. Curtis's *Bulletin* paper was, however, the most direct attempt to refute the basic assumptions on which Shapley's model was grounded. A lengthy section of Curtis's *Bulletin* paper was centred on Shapley's use of the Cepheid variables as distance indicators. After all, in 1917 Shapley had announced that in distance determinations the Cepheid variables were of much greater weight than the B-type stars and red giants because of the more definite knowledge of their dispersion in absolute magnitude. The B-type stars and red giants, he had written, 'can best be used as checks or secondary standards'.[100] Further, Curtis was not convinced that a period–luminosity relationship existed for the Cepheids. When he plotted the absolute magnitudes against the periods of the 37 Cepheids whose proper motions were catalogued (including the 11 used by Shapley) the resulting plot had the appearance of a scatter diagram.[101] Curtis did admit that if the Small Magellanic Cloud is not an exceptional region of space, then the existence of a period–luminosity relationship amongst the Cepheids of the Small Magellanic Cloud is, by analogy, the strongest reason for postulating a similar relationship among the Cepheid variables of our Galaxy.[102] But even this argument made him uneasy because at Washington he had remarked that the stars were so tightly packed in the globular clusters and Magellanic Clouds that they were in consequence special regions for variable stars.[103]

Curtis also dismissed Shapley's claim that if a single globular cluster was at the distance Shapley had derived, then his own contention that the clusters were at distances about one-tenth of those proposed by Shapley would be destroyed. 'While I hold to a theory of galactic dimensions approximately one-tenth of that supported by Shapley', Curtis noted, 'it does not follow that I maintain this ratio for any particular cluster distance. All that I have tried to do is to show that 10 000 light-years is a reasonable *average* cluster distance.'[104] Curtis preferred to secure distances to globular clusters by use of what he termed the 'average' star. Curtis was willing to accept correlations between large quantities of stellar data, but he felt that the diversity of the properties of the stars was too large to permit the use of limited amounts of data, especially when, as he believed was the case for the Cepheids' proper motions, the probable errors of the quantities observed were as large as the quantities themselves. Not enough was known about the giant stars for them to be employed as standard candles, and Curtis pointed out that in the solar neighbourhood the average absolute magnitude of the B-type stars was brighter than the average absolute magnitudes of the giants of other spectral types, yet this relationship was reversed in the clusters where the B-type stars were three magnitudes fainter than the giants of types K and M and about a magnitude fainter than those of type G.[105]

While Curtis conceded that Shapley's methods gave distances to the globular clusters that were relatively correct, he was still confident that Shapley's *absolute* distances to the globular clusters were about ten times too large. Consequently, he held as firmly as ever to his belief that the Galaxy is probably not more than 30 000 light years in diameter and perhaps 5000 light years in thickness.[106]

Some writers have mistakenly argued that Curtis defended the Kapteyn universe in the Great Debate.[107] Curtis was impressed by Kapteyn's techniques, but not by the data with which Kapteyn generated his results. In 1922 Curtis told a correspondent:

While I am ready to worship Kapteyn's *methods*, in which he has been fifty years ahead of the times, I can not, as most astronomers do, fall down and worship all the results which have come out of this mathematical mill.[108]

Then, to ram his point home he quoted, incorrectly, T. H. Huxley: 'Grinding peascods in a mill of super excellent grinding qualities will not produce wheaten flour'.[109] That is, since in his opinion Kapteyn's data were unreliable, even the most sophisticated mathematical techniques would not produce trustworthy results. Furthermore, in his *Bulletin* paper Curtis inclined towards a mixture of the spiral and the ring theories of the Milky Way. Shapley himself realised that Curtis 'swears by Newcomb and other patriarchs',[110] but *not* Kapteyn. In fact, Shapley actually claimed support for *his* scheme from Kapteyn's investigations.[111] He stressed that Kapteyn had found the density of stars along the galactic plane to be appreciable even at a distance of 40 000 light years from the Sun, and this indicated that the galactic system, even when the distant star clouds of the Milky Way were not taken into account, had a diameter about three times the value Curtis allowed as the maximum diameter of the Galaxy.

In addition to his criticisms of Shapley's chosen distance indicators, Curtis had another reason for attacking Shapley's model of the Galaxy: the defence of the island universe theory. If the spirals are as large as our galactic system, and if one assigns to them dimensions as great as 300 000 light years, then, Curtis admitted, 'the island universes must be placed at such enormous distances that it would be necessary to assign what seem impossibly great absolute magnitudes to the novae which have appeared in these objects'.[112] Curtis and many other supporters of the island universe theory had actually gone further and embraced a comparable-galaxy theory (that is, a theory in which the spirals are external galaxies comparable in size to the Galaxy). While Curtis commented that it is 'entirely possible to hold both to the island universe theory and to the belief in the greater dimensions for our galaxy by making the not improbable assumption that our own island universe, by chance, happens to be severalfold larger than the average',[113] he strove to avoid retreating to this position.

During the Debate and in his *Bulletin* paper Shapley hardly touched, explicitly at least, on the island universe theory. He was unrecognisable as the astronomer who had openly assailed the theory in 1918 and 1919. But Shapley still lacked a convincing alternative to the spirals as vast collections of stars, and his shifts of opinion in 1917, 1918 and 1919 as to their position with respect to the Galaxy reveal his uncertainty: in mid-

1917 he had contended that the spirals are distant star systems similar to the Milky Way; in 1918 he had placed the spirals, which he now believed were composed of nebulous matter, within the Galaxy; and in 1919 he had proposed that the spirals generally lay outside of the Galaxy but that a few of the nearer ones dash through the edges of the Galaxy running down stars and producing novae. In his *Bulletin* paper Shapley was scarcely more forthcoming about the spirals than he had been at Washington when he had tentatively suggested that the spiral nebulae are distant objects that lay outside of the Galaxy, although some of the nearer and brighter spirals appear, because of their distribution on the sky and their motions, to have some kind of relation to the Galaxy. A few months later Russell impressed on Shapley that the available evidence was not decisive for or against the island universe theory, and that it was a 'perfectly gratuitous guess'[114] that all island universes have to be of the same size. Shapley took Russell's prompting to heart since in his *Bulletin* paper he confined his criticisms to the comparable-galaxy theory. Furthermore:

Since the [comparable-galaxy] theory probably stands or falls with the hypothesis of a small galactic system, there is little point in discussing other material on the subject, especially in view of the recently measured rotations of spiral nebulae which appear fatal to such an interpretation.[115]

In 1920 van Maanen had once again measured the internal motions of a spiral nebula and the last section of this quotation refers to van Maanen's measures of the giant spiral M 33. In September van Maanen told Hale that he had completed a preliminary investigation of M 33, and that the motions were directed along the arms, and a little larger than those of M 101 that he had measured in 1916.[116] This was welcome support for Shapley's case, and in October 1920 he told Curtis he had not receded at all from his objections to the comparable-galaxy theory and that the 'rotation of Messier 33 and the others (which I believe you ignore completely) will not let me back down'.[117] A few days later van Maanen, who in April had discussed the spirals with Curtis, also informed Curtis of his measurements of M 33.[118] Curtis cautiously replied that if the motions of M 33 were to be confirmed then 'the result will be of the utmost theoretical value'[119] and in his *Bulletin* paper he conceded:

Should the results of the next quarter-century show *close agreement among different observers* to the effect that the annual motions of translation or rotation of the spirals equal or exceed 0″.01 in average value, it would seem that the island universe theory must be definitely abandoned.[120]

Curtis was still sure that such agreement would not be forthcoming.

Having scrutinised the tactics and arguments of the *Bulletin* papers of Curtis and Shapley we shall now ask how representative of astronomical opinion their analyses were. What seems likely from the available evidence is that at the time of the 'Debate' the island universe theory was the leading theory of the spiral nebulae. In July 1919, Shapley had told a correspondent that 'the majority of astronomers seem to believe that spiral nebulae *are* outside galaxies; they have not the most recent data on the subject, and I believe will not be hard to convince that spirals are nebulae'.[121] He had also remarked in 1919 that the island universe theory was a theory of long standing 'which at the present time has many adherents and appears to be growing in general acceptance'.[122] Little had occurred in the intervening year to alter this situation, and when we note that Shapley at this stage in his career was sometimes inclined to overestimate the support amongst astronomers for his own ideas, these statements are particularly telling.[123]

In addition to the many advocates of the island universe theory – astronomers such as Eddington, Jeans, Curtis, Campbell and Slipher – there were a substantial number who were still undecided as to whether or not the spirals were external galaxies. Amongst them was Hale. In 1916, van Maanen's measures of M 101 had allowed Hale to calculate a distance to the nebula. We will recall from Chapter 1 that the distance was 5000 pc, a figure that placed M 101 at the supposed boundaries of, if not beyond, the stellar system. Even then, he still had not declared his support for the island universe theory. Probably this was because, as he told Curtis, one could not base any final conclusions on the motions of spirals on the testimony of two plates.[124] In 1922, by which time van Maanen had measured further spirals with results that confirmed his previous studies, Hale was still warning that 'we are not yet certain whether [spiral nebulae] should be regarded as "island universes" or as subordinate to the stellar system'.[125]

One of the Lick astronomers who did not unreservedly accept the island universe theory was the double-star observer and its subsequent Director, Robert Aitken. He told a correspondent:

I would like to hear the debate between Curtis and Shapley. I have read Curtis's paper – a very good one – and have had long talks with Shapley also, and each one has many very good arguments to present. For my own part, I am still 'on the fence' on this question. I very greatly doubt the visibility of half-a-million or more 'island universes' on the one hand, and, on the other, I am not ready to accept Shapley's conclusions *on the basis of his measuring-rod*.[126]

Another old colleague of Curtis's at Lick, and the then Director of the Argentine National Observatory at Cordoba, C. D. Perrine, admitted to Curtis that he too was 'on the fence'[127] about the spirals. Perrine believed that the spectra of the spirals, and the novae in spirals, both favoured the island universe theory, while the systematic recession of the spirals counted against the theory.

The opponents of the spirals as island universes were not as confident of their position as the theory's advocates. Certainly by the time of the Debate Shapley had changed his mind about the status of the spiral nebulae three times in as many years. We have seen too that in his *Bulletin* paper Shapley had aimed his attack at the comparable-galaxy theory, not the existence of external galaxies generally. And Shapley's hypothesis that the spirals are genuine nebulae repelled from the Galaxy had also encountered severe difficulties. The repulsive force that Shapley suggested drove off the spirals was radiation pressure. In September 1920, after the Debate but before his and Curtis's *Bulletin* papers were finalised, Shapley reported to Hale: 'There will be no difficulty at all in driving off spirals; if their mean density is anything less than 10^{-10}, and they stop 1 % or more of the starlight falling on them, they can't fall into the system of stars. The mean density of the Andromeda Nebula is 10^{-21} or less, *unless* its mass is more than a million suns and its distance less than 1000 parsecs – both "unlesses" very unlikely'.[128] Hale replied that the 'radiation pressure results look promising, and should be followed up'.[129] Unfortunately for Shapley, Russell, who Shapley had urged to study the effect of radiation pressure on nebulous matter, had ruled against this mechanism a few days earlier. Russell told Shapley that even when he was 'liberal' towards the amount of radiation exerted by the Galaxy's stars on the supposed spirals, there was 'nothing doing!', since the gravitational pull exerted on the nebulae by the Galaxy was so much greater.[130] Russell then tried to account for the remarkable radial velocities of the spirals as a consequence of electrostatic forces, but he had concluded that these too could not begin to rival the attractive pull of the Galaxy's gravity.[131]

In the same letter in which he had presented these findings, Russell confessed to Shapley that he was perplexed by the spiral nebulae. He had estimated that Nova Aquilae, a nova that had flared in the Galaxy, had reached an absolute magnitude of -8. Yet if the novae in the spirals had absolute magnitudes of -8, the spirals had to be placed at the distances expected of island universes. Hence, Russell wondered:

...what about the novae in spirals and the island universe? And van Maanen's detection of motion in M 33 – of which he wrote me also. We are on the brink of a big discovery – or maybe a big paradox, until someone gets the right clue.[132]

Shapley was still trenchantly opposed to the existence of island universes, but he had little to offer in place of the view that spirals are galaxies. He answered Russell that 'I am as much baffled by spiral nebulae as you, except I see no reason for thinking them stellar *or* universes. What monstrous assumptions that requires before you get done with it.'[133]

During Shapley's stay at Mount Wilson his papers were edited by F. H. Seares. Seares agreed with Shapley that spiral nebulae were not island universes, and Shapley cited an investigation by Seares in his *Bulletin* paper as evidence against the comparable-galaxy theory. Seares had calculated that the surface brightness of our Galaxy is some seven to eight magnitudes fainter than the surface brightnesses of the spirals:[134] if the spirals are comparable in size with the Galaxy, and if they consist of stars, then the stars in them must be packed together far more closely than in the Galaxy. 'Alternative conclusions', Seares said, 'are obvious'.[135] The alternative conclusions were that the spirals are not composed of stars and are not comparable in size to the Galaxy. Instead, Seares seems to have believed that the spirals are genuinely nebulous.

Possibly the most determined opponent of the spirals as island universes was J. H. Reynolds. In 1920, in his paper entitled 'Photometric measures of the nuclei of some typical spiral nebulae', Reynolds had argued that there are two main types of spiral, and that they are distinguished by the appearance of their nebulosity.[136] That of the first type is of an amorphous, cloudy sort, while that of the second is condensed and granular. The granular condensations, Reynolds suggested, are probably stars in the process of formation. While he conceded that a spiral *may* end up as a galaxy, he emphasised that the number of condensations contained in a spiral was far fewer than the number of stars even in a globular cluster. Reynolds warned too that the nature of the novae in spirals was still open to doubt: no spectroscopic observations had yet been secured that would justify classing the spiral novae with the galactic novae. He also wanted more attention paid to the 1885 nova in the Andromeda Nebula since if that Nebula was a galaxy, then the nova 'must be credited with an intrinsic brilliancy far exceeding any of the galactic novae'.[137] Reynolds in addition provided the third piece of evidence – the other two being van Maanen's measures of internal motions and Seares's study of surface brightnesses – that Shapley used against the comparable-galaxy theory in his *Bulletin* paper. Reynolds had found that the outer regions of the spirals were bluer than the nuclei, and had hence decided that these distributions were not compatible with the spirals as island universes, he having assumed that a galaxy must be uniform in composition. Seares too had observed the colours of spiral nebulae, but maybe Shapley preferred to use Reynolds's study because he thought it too parochial to cite two sets of observations by Seares and one set by van Maanen, both fellow staff members at Mount Wilson, as major evidence against the island universe theory.

Weaving together all of these tenuous strands of evidence was, Reynolds cautioned, too risky. He proposed that on the question of the nature of the spirals astronomers should be neutral, and wait for more data before committing themselves to any particular theory.

In summary, then, we can state that at the time of the 'Debate' the island universe theory of the spiral nebulae, often identified as the comparable-galaxy theory, dominated its opposing theories. There was no credible alternative. Although Shapley still adhered to the hypothesis of the nearer spirals as clouds of nebulous matter that were repelled by the Galaxy, it had been all but wrecked by Russell's calculations. Moreover, the reflected-light hypothesis had been dropped in 1917 by Slipher, and in 1920 Reynolds too abandoned it. All that those astronomers who would not identify the spirals with island universes could do was to join Reynolds and point out anomalies in the theory of spirals as external galaxies, and press their colleagues to suspend judgement.

Notes

1 Clerke (1903) 542–3.
2 C. Young (1888) 511.
3 Newcomb (1906) 53.
4 These are rough distinctions and the variety among the main types was wide.
5 Paul (1976) Chapter 1. Paul's thesis is a most useful guide to contemporary papers on galactic astronomy in the late nineteenth and early twentieth centuries.
6 Pannekoek (1961) 473.
7 Seeliger (1898): for an analysis see Paul (1976) 75.
8 Paul (1976) 84.
9 On the lack of data on the distances of the stars see Nielsen (1963) 223–5.
10 Kapteyn (1908) 95.
11 Swihart (1968) 193, table 24.1.
12 The theme of what can be termed the 'Copernican tradition' – the desire to avoid placing the Sun in a privileged position – has been taken up in Seeley (1973), especially 229–30.

13 Shapley (1915*b*) 52.
14 H. Shapley to F. Moulton, 7 January 1916, Harvard.
15 J. Kapteyn to G. Hale, 23 September 1915, Hale.
16 Eddington (1914) 31.
17 Shapley (1915*b*) 86.
18 Curtis (1917*a*) 100.
19 H. Shapley to A. Eddington, 16 November 1916, Harvard.
20 Shapley (1919*a*) 314 and 315.
21 H. Shapley to H. Russell, 31 October 1917, Harvard. The argument that his data could not be reconciled with the hypothesis of the globular clusters as external galaxies had been presented to him in June 1917. Hertzsprung had told Shapley that he was now convinced that Shapley's counts of the number of stars in the globular clusters implied that the clusters could not be as large as the Galaxy (E. Hertzsprung to H. Shapley, 2 June 1917, Harvard). Indeed, after 1917 no one seriously suggested that the globular clusters were comparable-galaxies for another twelve years.
22 See, for example, Fath (1910).
23 H. Shapley to A. Eddington, 8 January 1918, Harvard.
24 H. Shapley to G. Hale, 19 January 1918, Harvard.
25 Eddington (1915) 376, and for his recollection of the reception he expected for this suggestion see Eddington (1930) 22.
26 Bohlin (1909) 20. Bohlin had speculated that the Galaxy had once been a part of a gigantic planetary nebula, a rotating shell of gas. The gas had condensed into different regions of space. Spiral nebulae, he wrote, are 'products of the broken parts of the shell, while the Milky Way . . . is derived from its equatorial belt'.
27 Gingerich (1975*a*) 347. In 1919 Shapley was to write that the galactic system 'may have originated in the combination of two clusters and has grown, as it appears to be growing now, by the accretion of other stellar systems – adding the smaller units such as globular clusters with ease, and the larger ones such as the Magellanic Clouds with some difficulty, if at all. It appears to be an example on a grand cosmic scale of survival of the fittest, that is, survival of the most massive and most stable' (Shapley (1920) 100).
28 Slipher had measured the radial velocities of a number of globular clusters. Their velocities were of the order of 100 km s^{-1} with no noticeable preference for velocities of recession or approach.
29 Shapley (1919*a*) 311.
30 Shapley (1919*b*) 266.
31 Shapley (1918*a*) 53.
32 Shapley (1919*b*).
33 We shall discuss Kostinsky's and Lampland's measures of the internal motions of spiral nebulae in Chapter 3.
34 Shapley (1919*b*) 265.
35 H. Shapley to H. Curtis, 1 October 1920, Allegheny.
36 On this scheme see Shapley (1920). Shapley, however, was not of the opinion that these nebulae would evolve into star systems.
37 H. Shapley to H. Russell, 31 March 1920, Harvard.
38 A. Eddington to H. Shapley, 24 October 1918, Harvard.
39 Russell (1918).
40 G. Hale to H. Shapley, 14 March 1918, Hale.

41 H. Russell to H. Shapley, 7 October 1918, Harvard: see also H. Russell
    to H. Shapley, 19 February 1919, 18 March 1919, and 20 March 1919,
    all Harvard.
42 G. Hale to H. Shapley, 14 March 1918, Hale. Shapley did try to obtain
    the proof that Hale desired. A few days before receiving Hale's letter
    he had told Jeans that 'I hope that you will be able to resume your study
    of the encounter of clusters . . . The relation of open and globular
    clusters to the greater galactic system shows the need of much further
    analytic work along that line' (H. Shapley to J. Jeans, 8 March 1919,
    Harvard). Also, amongst the Shapley papers at Harvard there are a
    number of mathematical notes that relate to Shapley's 'accretion'
    model of the Galaxy. These notes were later dated by Shapley as
    '1919±2'. But it is worth noting that even Hale sometimes found
    Shapley's notions too wild (see Gingerich (1973)).
43 J. Kapteyn to G. Hale, 23 September 1915, Hale.
44 G. Hale to J. Kapteyn, 4 November 1915, Hale.
45 G. Hale to J. Kapteyn, 17 July 1918, Hale.
46 G. Hale to A. Lawrence Lowell, 29 March 1920, quoted in Wright
    (1966) 326.
47 Kapteyn made this point to Shapley (J. Kapteyn to H. Shapley, 15 June
    1919, Harvard).
48 Schouten (1919a). In Shapley's opinion, Schouten's analysis was open
    to a number of grave objections. He doubted whether the luminosity
    distribution of the stars in the globular clusters could be accurately
    matched against that of the stars in the solar neighbourhood: it was
    likely that while a globular cluster's stars had all formed at about the
    same time, the heterogeneous mixture of stars around the Sun had
    been born at differing epochs. Nor did the observed luminosity curves
    for the clusters seem to agree with that derived for stars in the solar
    neighbourhood (see Shapley (1923a) 320–1).
49 Kapteyn & van Rhijn (1920). In 1919 Kapteyn had told Shapley that he
    could not accept the eccentric position of the Sun because the Milky
    Way is similar in all directions (J. Kapteyn to H. Shapley, 15 June 1919,
    Harvard).
50 Kapteyn (1922).
51 See, for example, Berendzen, Hart & Seeley (1976) 70–98.
52 Ravetz (1973) 266.
53 On Miss Leavitt see Bailey (1922).
54 Leavitt (1908) 107.
55 Pickering (1912) 1.
56 Pickering (1912) 2.
57 There are two possible reasons for this apparently puzzling inactivity.
    First, Miss Leavitt believed there was as yet no accurate way of
    calibrating her relation. In effect, she knew the slope of the curve of the
    p–l relation, but did not know where it fitted onto the luminosity axis.
    Nor did she explicitly state that these variables were identical to the
    galactic Cepheids. Nevertheless, she noted that the brighter Magellanic
    Cloud variables had light curves similar to one of the galactic Cepheids ·
    and she 'hoped . . . that the parallaxes of some variables of this type
    may be measured' (Pickering (1912) 3). The second reason is that the
    Director at Harvard was E. C. Pickering and his passion was the
    collection of data, rather than interpretation. Maybe because of this

attitude Leavitt did not want to venture too deeply into theory.
58 Hertzsprung (1913).
59 Russell (1913).
60 H. Shapley to E. Hertzsprung, 8 July 1913, Aarhus.
61 Shapley (1914) 449.
62 Russell & Shapley (1914) 431. Although this paper was co-authored by Shapley, it seems to have been mainly Russell's work.
63 Shapley (1918 *b*).
64 The calibrations of Hertzsprung and Shapley were assisted greatly by the Boss *Catalogue* of proper motions (*Preliminary general catalogue of 6188 stars for the epoch 1900*). Walter Baade, one of the foremost astronomers of a later generation, said in 1958 that 'the old Boss catalogue of proper motions of high precision . . . was just a fantastic achievement, which has never been surpassed' (Baade (1963) 3).
65 By doing so Shapley committed his famous 'error' of confusing the W Virginis stars and the Classical Cepheids: see Fernie (1969).
66 Shapley (1918*b*) 99.
67 Whitney (1972) 218.
68 Shapley and Adams also quarrelled over the First World War. Before the United States declared war on Germany in 1917, Shapley was not nearly as hard on the Germans as Adams would have liked.
69 W. Adams to G. Hale, 10 December 1917, Hale.
70 For a brief discussion of the method of spectroscopic parallaxes see Struve & Zebergs (1962) 201–2.
71 W. Adams to H. Russell, 4 February 1918, Princeton. For some years Adams did not agree that the brightest red stars Shapley was detecting in the globular clusters were giants. In September 1920 Shapley told Hale: 'I was much surprised, and pleased, by your remark that Adams had previously found from their spectra that the brightest red stars in clusters are giants. I knew nothing of that . . . . But I am greatly pleased to hear (and so are others who believe in the big Galaxy), that he now believes spectroscopic evidence proves these stars to be giants. It will have great weight in many obdurate corners. It should be, I think, the last straw that breaks the Campbell's back' (H. Shapley to G. Hale, 12 September 1920, Harvard). The Campbell referred to was of course the Lick Director.
72 H. Shapley to G. Monk, 28 January 1918, Harvard.
73 W. Adams to H. Shapley, 30 July 1921, Harvard.
74 H. Shapley to W. Adams, 13 August 1921, Harvard.
75 W. Adams to H. Shapley, 21 September 1921, Harvard.
76 This is clear from the Shapley–Adams correspondence in the Shapley papers at Harvard.
77 Hoskin (1976*b*).
78 C. Abbot to G. Hale, 3 January 1920, Archives of the National Academy of Sciences. A number of other topics were discussed as a subject for the evening lecture on 26 April, and at one time the Prince of Monaco was mentioned as a possible speaker on oceanography.
79 Shapley (1921) and Curtis (1921).
80 Hoskin (1976*b*) 169.
81 Hoskin (1976*b*) 175–81. The view of the Debate as a 'dramatic trial of strength' was begun in the 1920s, and by 1931 de Sitter was recalling how on 'April 26, 1920, took place the homeric fight on the scale of the

universe, the discussion between Shapley and Curtis' (de Sitter (1932*a*) 86).

82 Shapley (1969) 82.
83 H. Shapley to the Reverend J. Metcalf, 16 February 1918, Harvard.
84 H. Shapley to G. Hale, 22 February 1920, Hale.
85 W. Campbell to H. Shapley, 5 April 1916, Lick. Shapley replied: 'The nature of the Communications to the Proceedings of the National Academy of Sciences is such that full discussions of and references to earlier related work are not possible . . . I hope you do not believe that I have not appreciated the work at the Lick Observatory on Cepheid spectra or have tried to minimize its importance' (H. Shapley to W. Campbell, 12 April 1916, Harvard).
86 Shapley discussed the phrase 'Lick state of mind' in a letter to Hugo Benioff (H. Shapley to H. Benioff, 14 November 1923, Harvard).
87 H. Curtis to E. Barnard, 28 January 1920, E. E. Barnard Papers at the Joint Universities Library, Nashville.
88 G. Hale to W. Adams, 15 March 1919, Hale. This suggestion had actually been made to Hale via Merriam.
89 In fact, 'Russell made so substantial a contribution in support of Shapley that the question arose of whether he should be a third author of the published version [of the debate]' (Hoskin (1976*b*) 173).
90 H. Shapley to H. Russell, 31 March 1920, Harvard. Curtis had told Shapley that they should, metaphorically, shake hands before the encounter, but 'use our shillelahs' during the debate 'fo the best of our ability' (H. Curtis to H. Shapley, 26 February 1920, Harvard).
91 H. Shapley to H. Russell, 31 March 1920, Harvard.
92 Berendzen (1975) 72.
93 When it was suggested to Curtis and Shapley that they repeat the Debate for the summer meeting of the Astronomical Society of the Pacific, they both firmly rebuffed the idea (H. Shapley to H. Curtis, 9 June 1920, Harvard).
94 Shapley, as Russell reported to Hale, had performed poorly in the actual Debate (H. Russell to G. Hale, 13 June 1920, Hale).
95 H. Shapley to H. Russell, 31 March 1920, Harvard.
96 *Ibid.* In the months preceding the debate Curtis and Shapley had corresponded, but they had been concerned with the procedure of the debate and a very general discussion of their proposed talks.
97 Shapley (1921) 187.
98 A. Eddington to H. Shapley, 24 October 1918, Harvard. See also Eddington (1917*b*).
99 H. Shapley to A. Eddington, 19 November 1918, Harvard.
100 Shapley (1918*a*) 43.
101 Curtis (1921) 205.
102 Curtis (1921) 203.
103 Hoskin (1976*b*) 179.
104 Curtis (1921) 210.
105 Curtis (1921) 209.
106 Curtis (1921) 198.
107 For example, Berendzen, Hart and Seeley state that 'Curtis was primarily the spokesman for and defender of the Kapteynian scheme' (Berendzen, Hart & Seeley (1976) 39).
108 H. Curtis to Alter, 1 February 1922, Allegheny.

109 *Ibid*. Huxley had in fact written: 'Mathematics may be compared to a mill of exquisite workmanship, which grinds you stuff of any degree of fineness; but, nevertheless, what you get out depends on what you put in; and as the grandest mill in the world will not extract wheat-flour from peascod, so pages of formulae will not get a definite result out of loose data' (Huxley (1869) l). For the context of Huxley's remarks see Burchfield (1975).
110 H. Shapley to H. Russell, 31 March 1920, Harvard.
111 Shapley (1921) 173.
112 Curtis (1921) 210.
113 *Ibid*.
114 H. Russell to H. Shapley, 26 November 1920, Harvard. In 1919 Russell had called the spirals 'the most extraordinary objects in the heavens' (Russell (1919) 412).
115 Shapley (1921) 192.
116 A. van Maanen to G. Hale, 6 September 1920, Hale.
117 H. Shapley to H. Curtis, 24 October 1920, Harvard.
118 A. van Maanen to H. Curtis, 28 October 1920, Allegheny.
119 H. Curtis to A. van Maanen, 3 November 1920, Allegheny.
120 Curtis (1921) 214.
121 H. Shapley to A. Giberne, 19 July 1919, Harvard.
122 Shapley (1919*b*) 261.
123 For example, in March 1919 Shapley informed Charlier, an advocate of a small Galaxy, that his own 'results appear to be pretty generally accepted by English and American astronomers' (H. Shapley to C. Charlier, 20 March 1919, Harvard). Shapley's remarks beg the question of the opinions of Adams, the Lick school and many others.
124 G. Hale to H. Curtis, 8 February 1918, Hale.
125 Hale (1922) 12.
126 R. Aitken to E. Barnard, 5 January 1921, quoted by Hoskin (1976*b*) 174.
127 C. Perrine to H. Curtis, 5 January 1921, Allegheny.
128 H. Shapley to G. Hale, 22 September 1920, Harvard.
129 G. Hale to H. Shapley, 23 September 1920, Harvard.
130 H. Russell to H. Shapley, 17 September 1920, Harvard.
131 Russell suggested that three other possibilities remained to explain the high velocities: '(1) They are "primitive" – always thus. (2) They arise from the [gravitational] attraction of a huge system of which the Galaxy is a small isolated part. (3) They are not real motions, but a relativity effect of the general curvature of the [space time] "manifold"' (H. Russell to H. Shapley, 17 September 1920, Harvard). However, he admitted to having no strong inclination for any of these. A further discussion of his third suggestion is given in Chapter 5.
132 Russell's puzzlement over the novae did not last long. Two weeks later Russell told van Maanen: 'The nova in spiral nebulae must be rather a little fellow but this is not really surprising. I fancy that we find very little of the fainter galactic novae' (H. Russell to A. van Mannen, 5 October 1920, Princeton: quotations from the Russell papers are reproduced by permission of Princeton University Library). Thus he was now inclined to think that dwarf novae occur, but very few of this type had been detected in the Galaxy, whereas most of the novae observed in the spirals *were* dwarf novae.

133 H. Shapley to H. Russell, 30 September 1920, Harvard. The following month Shapley told Curtis that the radiation pressure calculations by Russell and himself had left the motions of the spirals more perplexing than ever (H. Shapley to H. Curtis, 24 October 1920, Harvard).
134 Seares (1920).
135 F. Seares to J. Kapteyn, 13 April 1920, Mount Wilson Archives.
136 Reynolds (1920a).
137 Reynolds (1920a) 753.

3

The Debate ends

We have seen that soon after its revival in the early 1910s the island universe theory assumed the leading role in the controversy over the nature of spiral nebulae, and that in 1920, at the time of the 'Great Debate', the theory was widely accepted. There was, however, another change of fortune ahead for the island universes. This was effected mainly by van Maanen who in the early 1920s measured several more spiral nebulae and found motions that seemed to refute the theory that they were external galaxies. One might have anticipated that these measures, because of their internal consistency, would have bowled over the supporters of the island universe theory. This supposition does not stand up to a detailed study, and we shall instead contend that during the early 1920s there were still many astronomers who believed that, on the whole, the theory was the best means of interpreting the properties of spiral nebulae. We shall then argue that Edwin Hubble's epoch-making discovery of Cepheids in spirals in 1923 and 1924 effectively ended the debate on the existence of external galaxies, and that these observations surprised the practitioners of nebular astronomy, but the demonstration they allowed of the extragalactic nature of spirals did not.

Internal motions

In 1922, Harold Spencer Jones, a future British Astronomer Royal, reported: 'Much controversy has raged as to the nature of spiral nebulae. The view held recently, when our galactic system was thought to be of much smaller dimensions than is indicated by more modern evidence, was that they were separate galactic systems.'[1] Spencer Jones, a past advocate of the island universe theory, had now rejected it because of the combination of van Maanen's measures of the internal motions of the spirals and Shapley's calculation of the size of the Galaxy. The contradiction between the existence of external galaxies and these two items of evidence had been emphasised by Shapley in 1919. As we have seen in Chapter 2, he had decided that the internal motions in M 101 detected by van Maanen, together with his own estimate of the Galaxy's dimensions, implied that the rotational velocity of the outlying parts of M 101 would be greater than the velocity of light, thereby reducing the island universe theory to absurdity.[2] After van Maanen had announced his preliminary measures of the giant spiral M 33, some astronomers began to query the comparable-galaxy theory, and, by association, the island universe theory. Now as long as van Maanen had firm results for only one spiral it was possible to avoid taking his measures seriously; when between 1920 and 1923 he carried the total of spiral nebulae that he had analysed to seven it was not.[3] The new measures were also being made on plates separated by longer periods than those of M 101 had been and so were of greater weight. Moreover, these extra measures agreed excellently with those he had made of M 101 in 1915 and 1916, and each of the seven spirals appeared to revolve – if the motions were interpreted as rotations – with a period of between 60 000 and 240 000 years.

But by 1921 van Maanen himself had become convinced that he was observing not pure rotations, but motions outwards along the spiral arms as Jeans's theories required. As we will recall from Chapter 1, Jeans's *Problems of cosmogony and stellar dynamics* had been received enthusiastically at Mount Wilson, and it was Jeans's mathematical investigations that van Maanen exploited to interpret his own observations.[4] In addition, as his researches progressed the more confident van Maanen became that his measures contradicted the island universe theory, whether they were viewed as rotations or motions along the spiral arms. In 1923 he concluded that the diameters of the spirals that he had measured ranged from a few light years to several hundred light years, and that their distances ranged from about one hundred to a few thousand light years.[5]

Also in 1923, van Maanen exhausted the supply of suitable early epoch plates and R. C. Hart, who has recently made a thorough examination of the internal motions, notes that it was now for the first time that van Maanen advanced detailed arguments to show that the measures were free of errors.[6] Van Maanen listed four errors that might conceivably be present:

(1) errors arising in the telescope systems;

(2)   magnitude errors;
(3)   errors arising from differences in quality or density between the old
and new photographic plates;
(4)   errors arising from the machines he employed to measure the
plates.

(1)   Van Maanen could not explain how consistent errors (if errors
there were) could have arisen from some peculiarity of the telescope
systems. The plates measured had, after all, been secured in three such
systems: at the 25- and 80-foot foci of the Mount Wilson 60-inch reflector, as well as the 18-foot focus of the Lick Crossley Reflector. He also
pointed out that if the telescope systems had generated the internal
motions, they had caused a left-handed 'rotation' for the four left-
handed spirals, and a right-handed 'rotation' for the three right-handed
spirals; a most unlikely outcome of such an error.
(2)   He accepted that the comparison stars were generally brighter than
the measured nebular points and that this might introduce magnitude
errors (of which more later). But he argued that a magnitude error could
produce 'only a bodily shift of the nebular points with respect to the comparison stars, or a radial shift [i.e. along the radii of the image field] due
to curvature of the [image] field and the smaller mean distance of the
nebular points from the center than of the comparison stars',[7] and not
the motion along the arms that he actually observed.
(3)   He discounted errors arising from quality changes in old and new
plates since he claimed that any differences were 'extremely small'.
Also, the exposure times of the new plates taken with the 60-inch had
been made 'so nearly equal to those of the old plates, that no difference
in the density can be seen'.[8]
(4)   Van Maanen noted that he had used a total of three machines to
measure the plates: a Zeiss stereocomparator, a stereocomparator built
at Mount Wilson, and an ordinary measuring machine fitted with an
extra microscope (to allow differential measures of two plates of M 101
as a check on the Zeiss stereocomparator). He wrote that it 'is clear that
defects in the optical system of the stereocomparator could never reveal
themselves as a rotatory motion of the nebular points, without equally
affecting the comparison stars. Moreover, the effect of such defects
would be eliminated in the reductions'.[9] Several other plates had been
examined in both stereocomparators, yet none of them had displayed
rotatory motions (as one would have expected if defects in the stereocomparators were producing errors). Van Maanen also stated that

another staff member at Mount Wilson, Seth Nicholson, had made sufficient measures of M 101, with both the Zeiss stereocomparator and the specially equipped standard measuring machine, to avoid any doubt as to the accuracy of the measures.

Many astronomers agreed with van Maanen. For example, in 1924 W. M. Smart of the University Observatory in Cambridge, England, checked van Maanen's method of reducing his measures, and Smart concluded:

I do not believe that anyone would be so bold as to question the authenticity of the internal motions – regarded either as rotational or as stream motion [along the arms] – found by van Maanen; in fact, the more one studies the measures, the greater is the admiration which they evoke.[10]

Furthermore, both van Maanen and Jeans claimed that the internal motions had been confirmed by others, and during the early 1920s this apparent agreement added to their credibility. In 1923 van Maanen stressed that Lampland, Kostinsky and Schouten had observed 'similar motions'[11] to those he himself had detected in M 51; but this assertion, as Hart has shown, does not resist detailed examination.[12] Indeed, even in the main 1916 paper on M 101 there is a clear inconsistency, as we shall now see. As a check on van Maanen's scrutiny of M 101, Nicholson had also measured the plates of the spiral. In van Maanen's opinion, the

	Nicholson*		van Maanen*
	(a)	(b)	
Proper motion of M 101 in right ascension	+0″.003		+0″.005
Proper motion of M 101 in declination	−0″.013		−0″.013
The radial component of the internal motion of M 101†	−0″.003	−0″.002	+0″.007
The rotational component of the internal motion of M 101	+0″.009	+0″.010	+0″.022

† For radial motions a − sign indicates a motion towards the centre of the nebula.
* Nicholson used 53 points of reference to get the results given in Column (a) and 42 for those in (b); van Maanen based his results on 87 points.

agreement of Nicholson's proper motions with his own was 'very satisfactory. Although Mr. Nicholson's value of the rotational component is smaller than [van Maanen's] a comparison with [van Maanen's] data shows that the difference is within the uncertainty of the determination.'[13] The proper motions and internal motions – divided into rotation and radial components – as determined by Nicholson and van Maanen are given in the table above.[14]

Notice that van Maanen's comment that Nicholson's rotation component was 'smaller' than his own blurs the fact that Nicholson's rotational value was *less than half* of van Maanen's. Nicholson had also found that the radial motion was directed *towards the centre* of the nebula, and not outwards as van Maanen had observed and as Jeans's theory of the spirals predicted. Van Maanen's claim was thus misleading. In 1919, Jeans had also emphasised the agreement of van Maanen's measures of M 101 with those made by others, and he wrote that Kostinsky had found 'similar results' in M 51 to those of van Maanen for M 101.[15] Yet in one part of M 51 Kostinsky had observed motions *towards* the centre of the spiral, whereas van Maanen's motions in M 101 were directed *away* from spiral's centre.[16] Thus Kostinsky's results could not be accurately termed as 'similar' to van Maanen's.

Hence both van Maanen and Jeans claimed a greater degree of corroboration for van Maanen's measures than the evidence justified. But in any case, van Maanen's investigations were to all appearances more meticulous and imposing than any other astronomer's. Van Maanen had examined seven spirals, and he had measured the motions of more points in the spirals: in M 51 Kostinsky had measured the motions of 36 points, Schouten a mere 9 points,[17] but van Maanen had traced the motions of 80 nebular points, and in his exhaustive study of M 33 van Maanen had actually measured the shifts of no fewer than 399 nebular points. Also, van Maanen had the reputation of a precise and thorough measurer of small displacements on photographic plates, and he was a staff member of one of the world's leading observatories. Therefore, even if astronomers had been aware of the differences between van Maanen's results and those of other measurers, it seems unlikely that they would have paid much heed to them.

What then was the status of the island universe theory in the early 1920s? In particular, is it true, as nearly all writers have since assumed, that van Maanen's direct measures of displacements within the spirals bowled over the theory's advocates? What of Spencer-Jones's contention that the combination of van Maanen's measures and Shapley's dimensions for the Galaxy had strongly influenced opinion against the island universes? This does not lend credibility to the hypothesis of the rejection of the theory simply because most astronomers did not subscribe to Shapley's model of the Galaxy. But evidence for such an hypothesis comes from correspondence between Shapley and van Maanen. In May 1921 van Maanen told Shapley that he had just finished measuring M 51: 'The motions look more convincing than M 101 or even M 33.' He went on: 'By this time Curtis and Lundmark must be the only strong (?) defenders of the island universe theory.'[18] Similarly, after Shapley had read one of van Maanen's papers for him at a meeting of the American Astronomical Society in August 1921, Shapley told him:

... your nebular motions are taken seriously now, and nobody but Very dared raise his head after I explained how dead the island universes are if your measures are accepted. And later he came round.[19]

When reading these remarks we must however remember that Shapley and van Maanen were close friends throughout Shapley's period at

Fig 12. Van Maanen's measures of internal motions in M 33. The arrows indicate the direction and magnitude of the annual motions. The length of the arrows represents the motions during an interval of about 2500 years. From van Maanen (1923*b*) Plate XIX.

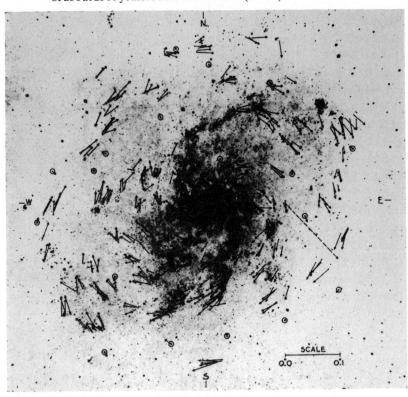

Mount Wilson and during the 1920s. The friendship was nourished in the early 1920s by Shapley's belief that he and van Maanen were together in the vanguard of astronomy. For example, in 1921 Shapley had congratulated van Maanen on his results: 'Between us we have put a crimp in the island universes, it seems – you in bringing the spirals in and I by pushing the Galaxy out.'[20] But Shapley's principal commitment in astronomy was to his own model of the Big Galaxy. He regarded his friend's measures as completely trustworthy, and he reasoned that as van Maanen's measures displayed real motions, the grounds for accepting his own galactic model were strengthened because the comparable-galaxy theory was exploded.

Support of the island universe theory

We see, then, that the most important evidence for the hypothesis that the island universe theory was rejected is unconvincing,[21] so let us now investigate the support for the theory in the early 1920s. It is significant that van Maanen's measures did not halt the publication of papers actively advocating the existence of external galaxies. A number of these contained derivations of the distances to the spirals from a study of novae. In 1922, for example, the available data for the novae in the Andromeda Nebula were analysed by Luplau-Janssen and Haarh.[22] They calculated the distance to the Nebula in two ways: (1) by comparing the mean maximum brightness of the novae in the Nebula – excepting S Andromedae which they regarded as unique – with the mean maximum brightness of the novae within the Galaxy; (2) by comparing the apparent distance apart of novae in the Nebula with the separation of novae within the Galaxy. Their answer of 100 000 pc, published in the widely read *Astronomische Nachrichten*, argued strongly for the Andromeda Nebula as an island universe.

Also in 1922, Ernst Öpik published 'An estimate of the distance of the Andromeda Nebula'. From the spectroscopically determined rotation of the Nebula, together with the assumption of a similar relation between mass and luminosity in the Nebula and the Galaxy, Öpik calculated the mass of the Nebula to be 1.8×10^9 solar masses, and its distance 450 000 pc. He concluded: 'the coincidence of results obtained by several independent methods increases the probability that this Nebula is a stellar universe, comparable with our Galaxy'.[23] One modern commentator has described Öpik's method as 'beautifully simple yet elegant',[24] and in 1936 Hubble termed it 'very ingenious'.[25] Öpik's paper was in addition brought to a wide readership by appearing in the *Astrophysical Journal*.

In 1922, D. B. McLaughlin, an astronomer at the University of Michigan, judged that the great weight of evidence favoured the island universe theory, although van Maanen's measures were anomalous and if they were proven to be genuine displacements, then the theory would have to be given up.[26] McLaughlin also used the formula van Maanen had given in 1916 to calculate the mass of M 101, and had thereby derived a result of about one-seventh of the usual estimate (10^9 solar masses) for the Galaxy's mass. Hence at the distance van Maanen adopted for M 101, the nebula should have exerted a powerful influence on the Galaxy. But van Maanen quickly countered McLaughlin's arguments. Persuaded by Jeans's researches and his own observations, van Maanen dismissed as worthless attempts to derive the mass of a spiral from its 'rotation': the internal motions disclosed motions along the arms, stream motions, not rotation.[27]

By his writings and lectures Jeans did much to bring van Maanen's measures to the attention of British astronomers. At a meeting of the Royal Astronomical Society in London in 1921 he had delivered a talk, illustrated with slides provided by van Maanen, 'On the internal motions in spiral nebulae'.[28] At the end of Jeans's discourse one questioner asked: 'What bearing have these measurements on the island universe theory?' Eddington answered that if they were to be definitely confirmed, then the theory would have to be relinquished. Jeans did not agree. Whereas Eddington was confident that Shapley's estimates of the dimensions of the Galaxy were essentially correct, Jeans was not. In fact, Jeans still favoured the Kapteyn model of the Galaxy, and this allowed him to embrace both van Maanen's measurements and the island universe theory.

Moreover, when in 1922 Jeans had analysed mathematically the internal motions and found that they did not conform to Newton's law of gravity, he had gone so far as to postulate a new 'generalized gravitational force which falls off approximately as r^{-1} and is not directed towards the nucleus [of the spiral]'.[29] The paper had been held for over a year, during the first part of which he had searched for a less revolutionary explanation of the motions (for example, by the effects of radiation pressure), and for the second part in an unsuccessful struggle to interpret the motions as a consequence of general relativity. While he wrote that the paper was put forward with 'all possible reserve', he was nevertheless prepared to compromise Newton's law of gravity rather than discard van Maanen's measures. We should however note that, as Jeans was fully aware, 'the temper of the times was highly suitable for such an apparently radical proposal, for it followed two of the most revolutionary decades in science. Unconventional proposals in physics were becoming commonplace.'[30] In 1923, for example, in an important and very subtle attempt to reconcile classical electrodynamics with the discontinuities of atomic physics, the eminent physicists Bohr, Kramers and Slater proposed that energy and momentum are not conserved in certain atomic interactions.[31]

One of the most enigmatic figures in twentieth century astronomy is Knut Lundmark.[32] He was often regarded as a brilliant astronomer, but Lundmark's researches on the spirals in the late 1910s and 1920s have been somewhat neglected by historians. We cannot afford to ignore them since Lundmark became one of the island universe theory's leading exponents. In 1920, he wrote a long paper entitled 'The relations of the globular clusters and spiral nebulae to the stellar system'.[33] When Lundmark sent copies to Curtis and Shapley, Curtis told him that it 'is a very valuable and excellent summing up of the evidence on these points',[34] and Shapley, while he disagreed with some of Lundmark's conclusions, commended the clear treatment.[35] With this paper Lundmark thrust himself to the very forefront of the debate on spirals and announced himself as a champion of the island universes.

In April 1921, when he journeyed from Sweden to the United States, Lundmark's involvement with the fortunes of the island universe theory deepened. He stayed in the United States for two years, spending lengthy periods at Lick and Mount Wilson.[36] After a few months at Lick he wrote a paper on the spiral nebula M 33 in which he claimed that part of the nebula could be resolved into stars.[37] Assuming that the 'resolved' stars were the brightest, he assigned to them the same absolute magnitude as the brightest stars in the Galaxy and thereby estimated the spiral's distance to be 330 000 pc. Furthermore, as Wolf had done in 1912, he compared the dark rifts in M 33 with what he hoped were similar structures within the Milky Way, and this procedure indicated to him that M 33 is about one hundred times as distant as the Milky Way rifts.

When in 1921 Lundmark composed his paper on M 33, van Maanen had already publicly announced his preliminary measures of that spiral.[38] Lundmark, however, did not mention them. This omission drew a letter in December 1921 from Shapley who asked Lundmark why he had ignored van Maanen's measures of M 33 and other spiral nebulae.[39] Lundmark's two replies to Shapley show that he was upset by the query. In the first, Lundmark, who was very much Shapley's junior as an astronomer (Shapley remember was now Director of the Harvard College

Observatory), pointed out that he had mentioned van Maanen's measures elsewhere and he had failed to do so again only because of the
limited space allowed for his paper in the *Publications of the Astronomical Society of the Pacific*. Lundmark wrote that although he had no definite objection to the reality of van Maanen's measures, he wondered
why there was no variation of the amount of measured outward motion
with distance from the centre of the spirals, as spectroscopic observations indicated there to be.[40] In the second letter, written the following
day, Lundmark was more forthright. As well as listing those papers in
which he had discussed van Maanen's measures, he recalled that in van
Maanen's recent paper on M 51 'the results of Dr. Schouten and
Kostinsky about the same object are, I will of course not say ignored, but
not mentioned . . .'.[41] A week later Shapley answered:

I fear that my inquiry disturbed you more than I intended. I thought possibly you
knew of some reason why van Maanen's measures of the spirals should be
gravely questioned. To find out if you had such a reason was the only point of my
inquiry. Apparently, you have no such misgivings; neither have I.

But even if these remarks were conciliatory, his next few comments were
unsparing:

Whether or not you care to recognise that his measures, if real, practically elim-
inate the 'island universe' hypothesis, which you seem to espouse at the present
time more strongly than any one, is not a matter I can properly concern myself
about.[42]

After this flurry of correspondence, no further letters were exchanged
between Shapley and Lundmark for several months. Then, in the April
1922 issue of the *Publications of the Astronomical Society of the Pacific*
(an important journal because the Lick, Mount Wilson and Lowell
astronomers often published there), a paper by Lundmark – advanced
clearly in response to Shapley's letter of the previous December –
appreared entitled 'On the motions of spirals'. Lundmark began by
declaring that it 'is obvious that the parallax for Messier 33 estimated by
van Maanen from the internal motions in this object is not in harmony
with the parallax value suggested by me as based upon certain spectroscopic results'.[43] He noted that the Mount Wilson astronomer F. G.
Pease had observed spectrographically an increase in the rotational velocity from the centre outwards in the Andromeda Nebula and the spiral
NGC 4594, but that van Maanen had observed motions in which the velocity near the centre was practically equal to the velocity in the outer

regions. Lundmark then showed that van Maanen had found the proper motions of the spirals to be smaller than their internal motions, and he discovered that the proper motions implied distances to the spirals some six to ten times larger than those indicated by the internal motions. In addition to this contradiction within van Maanen's data, Lundmark stressed that the methods of comparing the brightest stars – if stars they were – and novae in the spirals with those in the Galaxy, both gave distances larger than either the proper or the internal motions. Having criticised van Maanen's measures, Lundmark then turned to Shapley's theory of the Big Galaxy. Lundmark wrote that he was inclined to place the star clouds of the Milky Way a mere 12 000 light years from the Sun, and the open clusters at distances much smaller than those claimed by Shapley.

On 19 June an agitated Shapley told van Maanen that he had just read Lundmark's paper. Shapley viewed it as a serious assault on his Big Galaxy; he confided that 'I could talk to you (through this dictating machine) for an hour or so concerning the frailties of his discussion', and he dismissed Lundmark's remarks on the cluster distances as 'irrelevant'.[44] Two days later Shapley told a correspondent:

The fiercest attack on the scale of the universe is the one I did not see until the night before last, written by Lundmark. It seems to me that his work is full of half statements, of final conclusions based on some of his earlier suggestions ... I dislike, however, starting any controversy that might involve personal feelings.[45]

Shapley returned to Lundmark's claims two weeks later in a letter to van Maanen.[46] He wondered why Lundmark – who had now moved temporarily to Mount Wilson – insisted that he could discern stars in some spirals and that these stars had magnitudes of −6, but, at the same time, proposed that the brightest stars in the globular clusters were around magnitude 0, in violent opposition to the principle of the uniformity of nature. On 15 July Shapley complained about Lundmark's paper to Lundmark himself:

You make very many fine points in your discussion of the motions of spirals, but others that I would question. I am afraid that some of your statements on the scale of the Galaxy are misleading and throw doubt on my conclusions without justification. We could talk this matter over much better than we can write about it, so I will now only make a passing reference or two.[47]

In fact, these passing references lasted for over a page. Their main thrust was that Shapley doubted whether Lundmark was always complete in the treatment of points Lundmark judged to be against the Big Galaxy.

Shapley wrote that he did not want to incite antagonism and did not object to just criticisms of his work, but he believed that Lundmark's remarks on his galactic model were unfair. Then, to hammer his point home, Shapley added:

I heard Russell's statement concerning your paper, and I am sure you would not like to hear it yourself. And as you know, he admires your work in general.[48]

Lundmark did not reply to Shapley's letter until August 1922. The reason for the delay was, as van Maanen told Shapley, that Lundmark had been 'badly hurt by some of your remarks . . . However, he seems to be planning to postpone his judgement about you, till he has met you!'[49] The meeting between Shapley and Lundmark, which had been suggested a number of times in their correspondence, finally took place in April 1923.

During his stay at Mount Wilson, Lundmark, at the request of the acting Director, Adams, and van Maanen, had remeasured van Maanen's plates of M 33. In consequence, during his talk with Shapley, Lundmark disclosed that he had now altered his opinion of van Maanen's measures. A startled Shapley disclosed to van Maanen:

When Lundmark spent a few days here recently, I asked him pointedly what he thought of your proper motions in spiral nebulae. To my surprise he said he thought the best thing to do was to accept your measures at present. Apparently his work on Messier 33 convinced him that something was there. Hence, congratulations . . . Lundmark also suggested that he did not believe very heavily in the island universe theory, but one would never suspect that from his printed papers.[50]

Nor was Lundmark simply trying to appease Shapley: he recalled that for a time during his remeasurement of the plates of M 33, the 'situation seemed to be rather hopeless for the followers of the island universe theory'.[51] Here, then, was one of the theory's most vigorous advocates reluctantly admitting that van Maanen's measures exhibited real motions, and thereby conceding the dire consequences for the existence of external galaxies. However, by May 1924 Lundmark had again changed his mind. He had completed the reduction of the measures of M 33 and he was now of the opinion that he had not found any 'good agreement' with van Maanen's measures. Nor had he detected any systematic motion of the spiral.[52]

What lessons can be drawn from the Shapley–Lundmark correspondence of the early 1920s? First, Lundmark's 1922 paper 'On the motions of spirals' has to be read in the context of Lundmark's controversy with

Shapley. If Shapley had not provoked Lundmark by sending his letter of December 1921, it is unlikely that Lundmark would have fired off his attempted rebuttal of van Maanen's observations. Secondly, because of their different allegiances to the various techniques and methods of determining the distances of the spirals, Shapley and Lundmark were 'talking through' one another: Shapley trusted van Maanen's measures, whereas Lundmark placed most confidence in the novae as indicators of distances to the spirals. Thirdly, Shapley had ostensibly been drawn into the controversy with Lundmark because of Lundmark's attack on van Maanen's measures. Yet the controversy reached its peak, in Shapley's letter of 15 July 1922, after Lundmark had struck at Shapley's distance scale for the Galaxy; and the major reason for Shapley's adverse remarks was probably Shapley's desire to defend his own model of the Galaxy, a model that he believed was buttressed by van Maanen's measures. Fourthly, and most importantly, the controversy exemplifies how hard it was for an astronomer to question directly the reality of van Maanen's measures without his own arguments in turn being subjected to criticism.

There are sources other than published papers where we can find further implicit criticisms of the measured internal motions, as well as defences of the island universe theory. Eddington seems never to have accepted the internal motions as real. In 1917 he had warned against a premature acceptance of van Maanen's measures of M 101[53] and in *The mathematical theory of relativity*, completed in August 1922, he commented that it was usually supposed that the spirals were the most distant objects known, 'though this view is opposed by some authorities'[54] (these 'authorities' were surely van Maanen and his supporters). Slipher too was not swayed by van Maanen's measures. In 1921 the *New York Times* carried a story under the headline 'Dreyer Nebula No. 584 Inconceivably Distant' which possessed the appendage 'Dr. Slipher says the Celestial Speed Champion is "many millions of light years" away'.[55] Clearly Slipher still accepted the island universe theory,[56] as did Campbell. In 1921, Campbell told van Maanen that the strongest pieces of evidence for the theory were the dark lanes to be seen in edge-on spirals.[57] That is, Campbell was still of the opinion that the distribution of the spirals on the sky was explained by the presence of a ring of obscuring matter around our own Galaxy, and that the appearance of such matter in the edge-on spirals indicated they were external galaxies. Curtis never conceded that van Maanen had detected real motions, and in 1924 he confided to Slipher:

... my feeling is a mixed one of admiration for careful and honest measures on most difficult subjects ... and some measure of total disbelief that the motions he found exist at all in the quantities he gives.[58]

Although Curtis's own researches on the spirals had ended when he moved from Lick to Allegheny in 1920, during the 1920s astronomers sometimes wrote to him about them since he was still seen by many colleagues as the foremost champion of the island universes. For example, in 1924 S. M. Boothroyd of the Fuertes Observatory of Cornell University told Curtis that the island universe theory 'seems to me the only theory or rather the best theory to accept as a working hypothesis'.[59]

The opponents of the island universe theory, although they could form up behind van Maanen's measures to oppose the existence of comparable-galaxies (and usually by association, external galaxies), broke into disarray when the question of the composition of the spirals was fired at them. And in fact among the supporters of van Maanen's measures there were a number of radically different views about the nature of the spirals. In 1922, Shapley reaffirmed his belief that they are neither composed of stars nor comparable in size with the Galaxy. He wrote that the evidence favoured, though it did not establish, the hypothesis that the typical spirals represent a 'sidereal evolution not directly connected with that of stars'.[60] Shapley's hypothesis, however, gained few followers. In addition, in 1923 Shapley admitted to a correspondent:

I cannot give you the distances of the spiral nebulae, because we have no very good way of estimating them as yet ... Van Maanen's recent papers in the *Astrophysical Journal* on the Motions in Spiral Nebulae contain occasionally some estimates of the distances. I think these are the best we have, but Curtis and others would place the spirals at much greater distances.

I generally say, in a discussion of the subject, that the larger spirals are probably at about the same distance as the larger globular clusters – that is, between five and fifty thousand light years.[61]

We should also note that some astronomers, including van Maanen and Reynolds, advocated the hypothesis that, although not external galaxies, the spirals are star producing objects. Reynolds, a loyal defender of van Maanen's measures, believed the internal motions had been underpinned by Jeans's theoretical researches. Reynolds, nonetheless, was not fully convinced by Jeans's scheme, and in 1923 he toyed once more with the reflection hypothesis of the spirals.[62] In 1923, F. A. Lindemann, an Oxford professor of physics who was later to find fame as a scientific adviser to Winston Churchill, revived the hypothesis that the spirals are galactic in origin, are composed of dust, and have been expelled from the galactic plane by radiation pressure. Lindemann argued that van Maanen's measurements demonstrated that the spirals were not, as he wrote was often supposed, island universes.[63] Although his proposals generated a heated discussion, they won very little support.[64]

In 1923, then, there was much uncertainty over the nature of, and distances to, the spiral nebulae. The split between the supporters and opponents of the island universe theory was merely one manifestation of this confusion. Lindemann's ideas, and the reactions of astronomers to them, was another, as was Jeans's proposed non-central force law of $r^{-\frac{1}{2}}$ to explain the internal motions of the spirals. Furthermore, in 1923 even Shapley, one of the most strident opponents of the island universe theory, was conceding that a visible system of stars lay well outside of the Galaxy. In that year both Edwin Hubble, a Mount Wilson astronomer, and Shapley were studying the star cloud NGC 6822. By July 1923 Hubble had found five variables in the cloud, and he asked if Shapley could investigate the object at Harvard.[65] Shortly afterwards Shapley estimated its distance from a comparison of NGC 6822's dimensions with those of the Large Magellanic Cloud, and a comparison of the magnitudes of the Galaxy's supergiant stars with what he expected were similar stars within the star cloud. His answer was of the order of a million light years, and so placed NGC 6822 far beyond the limits of the galactic system.[66] At the same time, Shapley was finding that van Maanen's measures, as interpreted by Jeans, were not in harmony with nebular distribution and nebular brightness: if the spirals were producing star clusters in their outer regions, as Jeans proposed, there should have been many more giant stars in the vicinity of the spirals than Shapley found was the case.[67]

Hubble and Cepheids

In late 1923, amidst the prevalent and growing uncertainty about the nature and distances of spiral nebulae, one of the most momentous discoveries in the history of astronomy was made: Edwin P. Hubble's detection of a Cepheid variable in the Andromeda Nebula. As we shall see, Hubble was now set on a course that would soon bring the long-standing debate on the existence of external galaxies to a swift end.

To begin our account of Hubble's find and its reception, we shall briefly consider Hubble's early career and the concerns that brought him to study spiral nebulae.[68] Hubble was an undergraduate at the University of Chicago (working as a laboratory assistant to the eminent physicist Robert A. Millikan in his junior year), after which he studied Roman and English law at Oxford. He even practised law for a year, but in 1914 he gave up his legal practice to pursue postgraduate research on faint nebulae under E. B. Frost at Yerkes Observatory. Hubble's legal training, nevertheless, seems to have affected the manner of his science because one of his traits was a meticulous sifting of evidence. In 1917 Hubble completed his dissertation and immediately enlisted in the American army to fight in Europe. After leaving the army in 1919, he joined the Mount Wilson staff (the post having been offered to him in 1917 when Hale was looking for astronomers to use the soon to be completed 100-inch reflector), and continued the investigation of nebulae that he had begun at Yerkes.

During his first years at Mount Wilson, Hubble performed what other astronomers viewed as excellent research on reflection nebulae,[69] but he was soon embarking on a study of the spirals. From his writings before, during, and after 1923, it seems that Hubble always favoured the island universe theory. In 1917, principally because of the perplexingly high radial velocities of the spirals, he had argued that they are island universes.[70] In 1922, tucked away in a paper on 'The source of luminosity of galactic nebulae', he had estimated the distance to the spiral M 33.[71] By comparing four stars in what he identified as an emission nebula of M 33 with similar stars in the Galaxy, he calculated the spiral's distance to be 30 000 pc. Hubble had thus placed M 33 about an order of magnitude further away than did his colleague van Maanen, and he had also assumed that M 33 contained stars similar to those within the Galaxy. Maybe Hubble did not emphasise his answer of 30 000 pc precisely because it disagreed with distances derived from van Maanen's measures: even after detecting numerous Cepheids in spirals, Hubble, as we shall see, hesitated to announce publicly his results because of this discrepancy.

As we saw in Chapter 1, Ritchey had concluded that the images of the condensations in large spirals were softer than equally faint images on photographs of star fields. Hubble decided to investigate these condensations with the new 100-inch telescope in two ways: first, he took short exposures centred on the nuclear regions, and secondly he secured longer exposures centred both on the outer regions of the spirals and neighbouring selected areas. In this way he hoped to decide whether the non-stellar appearance of Ritchey's images was compatible with them being stars. The condensations in the nuclear regions were superimposed on relatively dense, unresolved regions of the spirals, but short exposures of the central regions would not be so affected. The condensations in the outer regions, which were not superimposed on the nebulosity, were subject to off-axis aberrations of the telescope but they could be avoided by centering on the outer regions. On these plates he detected condensations with the smallest angular diameters yet recorded. As Hubble recalled in 1936: 'They fully established the essentially stellar appearance of the photographic images of the great majority of the condensations when effects of photographic smearing were avoided.'[72] This did not mean that Hubble was fully convinced he could see individual stars; rather, the images were indistinguishable from stellar images but until they could definitely be shown to be stars by the display of some stellar characteristics, the case for the island universes would not be much strengthened. For example, when in 1922 Hubble proposed that stars were visible in the outer regions of a giant member of the spiral family, M 87, he justified this claim to a doubtful Shapley by writing,

My own preference concerning the objects clustering around M 87 is to call them stars until they may be definitely shown not to be stars. I believe they are stars but would not care to argue the point or to use this conclusion as a basis for

Fig 13. The 100-inch Hooker telescope at Mount Wilson (Courtesy of Mount Wilson and Las Campanas Observatories, Carnegie Institution of Washington).

speculation until better plates have been obtained. The stars begin where the nebulosity stops.[73]

'I remain sceptical as to the stellar nature of the small objects around M 87', Shapley answered. 'I see, however, that the data can be considered as insufficient to show that they are nebulous spots.'[74] Hubble was impressed by the agreement between the forms of nebulae and those predicted by Jeans's theories on the development of spiral nebulae. So, in addition to his long standing support for the island universe theory, he had an additional reason to suspect that the observed condensations were stars since Jeans had proposed that giant nebulous spirals (such as he believed M 87 to be) gradually develop spiral arms by the transformation of the material in their outer regions into stars.

In late 1923 Hubble began to focus on a study of novae in the spiral nebulae. In so doing, he took numerous plates of the Andromeda Nebula. For someone anxious to establish an accurate distance to the Nebula this programme had much to recommend it. In particular, by observing more novae he would be able to determine with greater precision the mean apparent brightness of novae in the Nebula, and thus estimate its distance with more certainty.[75] He was also excellently placed to pursue such a programme because he had access to the world's most powerful telescope, the Mount Wilson 100-inch. Whatever his actual motives, he was soon richly rewarded.

On 19 February 1924 Hubble disclosed to Shapley:

You will be interested to hear that I have found a Cepheid variable in the Andromeda Nebula (M 31). I have followed the Nebula this season as closely as the weather permitted and in the last five months have netted nine novae and two variables . . . The two variables were found last week.[76]

The Cepheid had first revealed itself on a plate taken in October 1923. Initially Hubble had assumed it to be a nova, but by checking earlier plates and drawing a light curve he had quickly realised that it displayed the characteristics of a Cepheid variable and he had soon estimated an approximate period.[77] By making the fundamental assumption that throughout the universe the Cepheids with the same periods have the same absolute magnitudes, knowing the star's period, and exploiting Shapley's period–luminosity curve, he had determined the distance to the Andromeda Nebula to be 'something over 300 000 parsecs'. Hubble, however, warned that 'if the star were dimmed materially by shining through nebulosity the distance would be correspondingly reduced'. The data on the second variable were unclear, but Hubble suspected it too was a Cepheid. 'I have a feeling', Hubble added, 'that more variables will be found by careful examination of long exposures. Altogether the next season should be a merry one and will be met with due form and ceremony.' Hubble was not disappointed. On 25 August he reported to Shapley that he had found 16 variables in or in front of the Andromeda Nebula, of which three were definitely Cepheids. Hubble added that he had detected 15 variables in M 33, and that he was inclined to think that he had observed variables in M 81 and M 101.[78] By September he could confirm that the Andromeda Nebula 'shows a definite period luminosity relation among the ten variables for which periods have been determined'.[79] Although a few types of stars were recognised to be intrinsically brighter than even long-period Cepheids, the Cepheids exhibited brightness variations that made identification of them relatively straightforward. Here, then, was the conclusive proof he had desired of the stellar nature of some of the condensations in spiral nebulae and he was

Fig 14. The light curve drawn by Hubble for the first Cepheid to be discovered in the Andromeda Nebula. This drawing accompanied Hubble's letter of 19 February 1924 to Harlow Shapley.

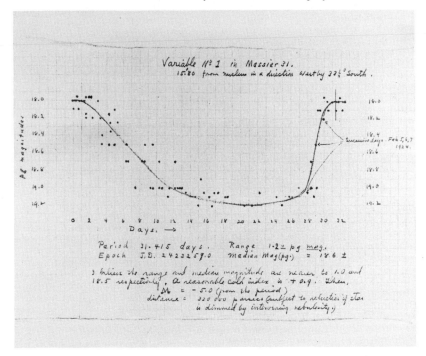

Fig 15. Hubble on Mount Wilson in 1923 (Courtesy of the Archives, California Institute of Technology).

now sure that he had resolved the outer regions of M 31 and M 33 into clouds of stars.

Although some heard of Hubble's observations of Cepheids in spirals shortly after they were made, they were not formally laid before astronomers until 1 January 1925 when a paper was read by H. N. Russell on Hubble's behalf to a joint meeting of the American Astronomical Society and American Association for the Advancement of Science in Washington D.C.[80] Russell announced that Hubble had definitely detected 12 Cepheids in the Andromeda Nebula and 22 Cepheids in M 33, and that from the Cepheid *p–l* curve Hubble had derived a distance to each of the spirals of roughly 285 000 pc, placing them both far outside even Shapley's model of the Galaxy. This result, nevertheless, rested on three assumptions. First, that the Cepheids were in fact connected with the spirals. Secondly, that there was no serious amount of absorption due to nebulosity within the spirals. Thirdly, that a Cepheid in one star system does not have a different intrinsic luminosity to one with the same period in another star system. The first assumption seemed highly plausible, especially as no Cepheids had been detected on several plates of neighbouring areas of the sky. Nor was there any evidence that the second assumption was seriously in error, while the third relied on the principle of the uniformity of nature.

Hubble's Cepheids obviously supported the island universe theory, while van Maanen's internal motions just as clearly opposed the theory. Astronomers were thus faced with three possibilities: (1) the reconciliation of van Maanen's measures with Hubble's observations; (2) the rejection of the Cepheid interpretation of Hubble's observations; (3) that van Maanen's measures were gross exaggerations of the actual motions. Most astronomers rapidly plumped for choice (3). Certainly within a year or two of the first Cepheid's detection the debate on the existence of island universes was effectively over. The Cepheids, as we shall see later, were viewed by opponents of the island universe theory as distance indicators accurate enough to establish the spirals as external galaxies, and corroboration of this interpretation was provided by Hubble's unambiguous resolution of the outer parts of the Andromeda Nebula and M 33 into stars.

But not all astronomers accepted the lessons of Hubble's observations immediately. For them the issue was not clear, and for some the sense of confusion over spiral nebulae was deepened by what they saw as two seemingly reliable, yet blatantly contradictory, sets of observations. For example, in March 1924 Paul Merrill, a member of the Mount Wilson staff and a highly talented spectroscopist, told Curtis that Hubble had found a Cepheid in the Andromeda Nebula whose period led to a distance of 750 000 light years. The bewildered Merrill then added:

The applicability of the [*p–l*] relation is, of course, in doubt but probably no more so than in some other cases where it has been applied. What do you think of van Maanen's measurements? It seems difficult to disbelieve the reality of the outward (i.e. along arms) motions which he had found in seven objects but they are very puzzling, as Jeans's analysis has clearly shown. Can relativity provide a region of space where circular motion can take place without the necessity of centripetal force?[81]

Hubble was fully aware of the conflict between the implications of van Maanen's internal motions and his own observations of the Cepheids, and the reason for Hubble's reluctance to publish his results earlier was, as he disclosed to Russell in February 1925, this 'flat contradiction'.[82] Hubble, we should recall, was still a relatively junior member of the Mount Wilson staff, while van Maanen had worked at the Observatory for twelve years. Indeed Hubble, by temperament careful not to jump to conclusions, initially avoided committing himself to any particular interpretation of his observations. But in August 1924, when he was convinced of the existence of 16 variables – three of which he said were definitely Cepheids – 'in or in front of Messier 31' and 15 variables in M 33, he confided to Shapley:

I feel it is still premature to base conclusions on these variables in spirals, but the straws are all pointing in one direction and it will do no harm to begin considering the various possibilities involved.[83]

By early 1925, at the latest, Hubble had become convinced that van Maanen's measures were spurious. This is shown plainly in a manuscript by Hubble entitled 'Notes for a talk on the application of the Cepheid criterion to spiral nebulae'.[84] These notes can be dated to late 1924 or early 1925, and Hubble probably delivered a talk based on them to the weekly Astronomy and Physics Club at Mount Wilson. Hubble emphasised the reliability of the period–luminosity relationship, and that as well as constructing period–luminosity curves for the Andromeda Nebula and M 33, he had also established that the relation existed for the Cepheids in the star system NGC 6822. Hubble in addition contended that 'Mr van Maanen's results are quite isolated' not only from other sorts of observations, but even from those of the other researchers who had examined the internal motions of spiral nebulae. In Hubble's opinion, only Schouten's analysis of M 51 deserved any consideration, but Schouten had observed motion in the *opposite direction* to that detected by van Maanen,[85] and Hubble bluntly insisted:

The real controversy is between the universality of Cepheid variation and the validity of Mr van Maanen's motions.

By the end of 1925 nearly all astronomers had considered this choice and had decided, or would soon decide, that van Maanen's measures had to be rejected.

To help us understand this shift of opinion we shall now investigate the reaction to Hubble's Cepheid observations of the leading advocates, apart from van Maanen himself, of the reliability of van Maanen's measures: Shapley, Russell, Jeans and Smart. In answer to Hubble's letter of 19 February 1924 (quoted on p. 114), Shapley wrote that the letter was 'the most entertaining piece of literature I have seen for a long time'.[86] The rest of Shapley's reply does not explicitly indicate whether or not he accepted that Hubble had found a Cepheid. There is one passage, however, which reads as if it might be a politely veiled suggestion that he did not:

Incidentally, your experiences have probably shown you that in the denser part of the nebulosity every star is a variable, depending upon exposure time, development, and other factors. We have here some ... [false] variables in the middle of clusters, with a range of one or two magnitudes. The distance of your variable from the nucleus and the lovely number of plates you have now on hand of course assures you of genuine variability for these stars.

Shapley pointed out too the difficulties of accurately estimating the magnitude of the variable, and that his experience of Cepheid variables with periods greater than twenty days 'is that they are generally not dependable – they are likely to fall off of the period–luminosity curve. Possibly they are transition types one way or other from U Geminorum variables, or the RV Tauri type. Nevertheless, they appear so far to be less than two magnitudes fainter than the values predicted by the period–luminosity curve.' These remarks are important because if the variable was indeed absolutely fainter than Hubble had assumed, then its distance would have to be reduced accordingly. Shapley met with Hubble in April 1924, and by September any doubts that Shapley may have harboured had vanished. He now admitted to Hubble that 'I do not know whether I am sorry or glad to see this break in the nebular problem. Perhaps both. Sorry because of the significance for the measured angular rotation, and glad to have something definite and interesting come to hand.'[87] And seven months later Shapley told a correspondent:

The new Mount Wilson work is, I think, pretty conclusive that the spiral nebulae are made in part of stars – that is, that the brighter spirals are stellar. This conclusion is, of course, in flagrant disagreement with van Maanen's measures of angular motions. These angular motions seemed so very definite and general, that I, along with the rest who have seen his plates and know the character of his measuring skill, considered the problem settled. But others have not been able to support van Maanen very conclusively. And now Hubble's measurement of the periods and apparent magnitudes of many Cepheids in Messier 33 and Messier 31 is of a quite higher order of dependability.[88]

Russell also lost faith in the correctness of van Maanen's measures very quickly. Early in 1924 he had confidently remarked that the larger spirals were only 10 000 light years away and that 'everyone is now agreed that these great spiral nebulae are actually objects in which there is an outward expanding motion into and along these faint arms, a rotatory and outward motion'.[89] But in December 1924 Russell sent to Hubble his heartiest congratulations on the detection of Cepheids in spiral nebulae: 'They are certainly quite convincing. I heard something about them from Jeans a month or two ago.'[90] In October 1924, Jeans, probably van Maanen's most influential supporter, had concluded that the measured motions could not be real.[91] Jeans, who had heard about Hubble's observations during a visit to Mount Wilson in the summer of 1924, told Russell that he had performed some calculations that had removed his earlier suspicions of the period–luminosity relationship; in consequence, 'I fear van Maanen's measurements have to go'.[92] Reynolds had defended van Maanen's measures from the first, but in October 1924 Jeans told him of Hubble's researches.[93] In November Hubble himself described his latest findings to Reynolds (although he may have related his first results to Reynolds earlier in the year).[94] The following month Reynolds conceded that 'the great massive spirals M 31 and M 33 are possibly comparable in size and mass with the galactic system' although he insisted that the great majority of spirals are intrinsically small in comparison.[95] Here, then, was a switch by Reynolds to a restricted version of the island universe theory. As to van Maanen's measures, another 'twenty five years will see these measures confirmed or otherwise'.[96] By 1926 Reynolds had admitted that M 31 and M 33 were at the distances Hubble claimed, although he still did not accept that every spiral was an island universe.[97] In 1924, W. M. Smart had checked van Maanen's method of reducing his measures and had confirmed that they exhibited genuine motions. But in 1928 Smart wrote that the Universe is a vast assemblage of galaxies, and he now judged that either the internal motions are 'fictitious in the sense that they are very many times too large or else the estimates of the distances of the nebulae are erroneous. The latter alternative involves the accuracy of the period–luminosity law for Cepheid variables, but the consensus of contemporary astronomical opinion is against throwing the blame in this direction. The discordance can only be cleared up in the future when a sufficiently long interval has elapsed to allow the attainment of much greater accuracy in the proper motions.'[98]

Once van Maanen's measures had been rejected the case against the island universes collapsed under the overwhelming weight of evidence in their favour.[99] Yet why did it take astonomers until 1924 to realise that the Cepheids were capable of resolving the island universe debate? *Prima facie*, there seem to be three possible answers to this question:

(1)   Astronomers did not believe Cepheids were to be found in spiral nebulae.

(2)   Those who did believe that Cepheids were present in spirals, and who accepted the reliability of the p–l relation, imagined them to be too faint to be discerned against the nebulous background of a spiral nebula.

(3)   Astronomers did not realise, or accept, the importance of the Cepheids as distance indicators. In particular, they might not have accepted that *all* Cepheids in the universe of a particular period have the same intrinsic luminosity.

Possibility (1) can be ruled out immediately. Even in the early 1920s, when van Maanen's measures of internal motions did influence astronomical opinion, many astronomers viewed the spirals as island universes, and if the spiral nebulae were galaxies of stars, then, by the principle of the uniformity of nature, they should, like our Galaxy, contain Cepheid variables. Nevertheless, the fact remains that before Hubble's discovery was formally announced only one astronomer – Knut Lundmark – explicitly suggested searching for Cepheids. When in September 1924 Lundmark addressed the Astronomische Gesellschaft at Leipzig, he noted that a few claimed variable stars had already been observed in spiral nebulae and argued that if Cepheids could be found, then a new method of securing distances would be available.[100] But when Lundmark made this suggestion, Hubble, unknown to Lundmark, had already detected Cepheids in the Andromeda Nebula. Writing in 1924, Lundmark himself picked possibility (3) to explain why Cepheids had not already been observed in spirals:

We see that the Novae at their maximum are among the absolutely brightest stars in the Galaxy. If the same is the case with the Novae in spirals it may explain why *Cepheids* or other variable stars have not been discovered yet. These stars being absolutely fainter than the Novae will only reach 18[th magnitude] or 19[th magnitude] at their maximum, and will not be easily discernible on the nebulous background.[101]

There is, however, one puzzling point: either Lundmark wrote these words before his Leipzig address of 1924, or he had forgotten that variables had been observed in a spiral, a fact to which he had referred at Leipzig. The latter explanation seems so implausible that the former alternative must be favoured. In fact, a few years earlier J. C. Duncan and Max Wolf, independently of each other, had discovered variables in the giant spiral M 33.[102] In 1920, while searching for novae on Mount Wilson plates, Duncan noticed three faint variables. By mid-1922 he had followed their brightness variations on 17 plates, but he had not established if they possessed fixed periods. In 1922, Wolf at Heidelberg too found a variable in M 33, but neither Duncan nor Wolf had suggested that any of the variables might be Cepheids. With the benefit of hindsight, one might have expected the discovery of these variables to have produced eager attempts to detect more, as had happened with the detection of novae in 1917, but this was not the case.

Can possibility (3), a widespread distrust of the Cepheid criterion, explain why it was not until 1924 that Cepheids were actively searched for in spirals? In the early 1920s, Shapley realised that his own model of the Galaxy was under assault because of the criticisms of the calibration, and sometimes even the existence, of the period–luminosity relation.[103] In 1922, in an attack on Shapley's use of the Cepheids as distance indicators and thereby a defence of their own techniques for elucidating the structure of the stellar system, Kapteyn and van Rhijn analysed the proper motions of those Cepheids for which they had data. Kapteyn and van Rhijn argued that the short-period Cepheids (essentially the cluster variables) and the long-period Cepheids were fundamentally different.[104] The short-period Cepheids, they pointed out, are evenly distributed over the sky and possess large proper motions. Further, their own calculations indicated that the short-period Cepheids were not giant stars as Shapley believed; thus Shapley had exaggerated their brightness. In consequence, the distances – and hence the dimensions of the Galaxy – that Shapley had estimated with the aid of the Cepheids were too large by a factor of about seven.[105]

Shapley was sure that Kapteyn and van Rhijn had blundered; the large proper motions of the cluster type variables were not due to nearness as Kapteyn and van Rhijn thought, but to the large space velocities of the stars.[106] Shapley also announced that he had observed 13 variables in the

Small Magellanic Cloud that exhibited the light curves of typical cluster variables and that sat perfectly on his period–luminosity curve. Thus because all of the stars in the Small Magellanic Clouds are at roughly the same distance, these 13 variables refuted the claim by Kapteyn and van Rhijn that the short-period and long-period Cepheids are intrinsically different. Such 'a direct extension of the observed period–luminosity curve to cluster type variables in the Small Magellanic Cloud itself would appear', Shapley declared, 'to be specially decisive'.[107]

However, as J. D. Fernie, a leading authority on Cepheid variables, noted in 1969,[108] the periods that Shapley quoted for the 13 stars were badly wrong, as a comparison of Shapley's values with the 1966 values of C. Payne-Gaposchkin and S. Gaposchkin reveals:[109]

Star (Harvard number)	Period (1966) (days)	Shapley's Period (days)
1408	1.63	0.715
1420	1.97	0.67
1634	3.97	0.81
1655	1.61	0.444
1731	2.14	0.65
1739	1.81	0.63
1741	eclipsing variable	0.40
1823	3.35	0.795
1928	1.66	0.608
1943	1.62	0.81
1964	2.09	0.687
2002	2.34	0.702
3610	not measured	0.60

Shapley was anxious to rebut Kapteyn and van Rhijn's arguments. He had measured the magnitudes of cluster variables in the Small Magellanic Cloud and so 'knew' – because of the period–luminosity relation – the period that each star should possess if it were a cluster variable. Shapley duly seems to have convinced himself that the 13 stars had the expected periods. Probably by 'knowing' the periods of the stars he drew the light curves incorrectly.[110] Incorrect light curves or not, astronomers surely would have accepted that Shapley had buttressed his argument that a single relationship linked the cluster variables and the long-period Cepheids and so strengthened the foundations of his calibration of the period–luminosity relationship.

Additional support for Shapley's calibration was provided by the analysis by R. E. Wilson, an established American astronomer, of the proper motions and radial velocities of 74 Cepheids and 10 cluster variables within the Galaxy.[111] Wilson concluded that the zero-point of Shapley's curve needed shifting by an amount that would reduce the distances derived from the relation by about one-fifth. Shapley's surviving letters from 1923 disclose that he viewed Wilson's 'confirmation' of his calibration to be very important. For example, he told a correspondent that Wilson had showed 'pretty definitely that my distances cannot be more than thirty per cent too large. This is considerably different from Lundmark's factor of 3.0, or Schouten's factor of 7.6.'[112] Despite Shapley's criticisms of Kapteyn and van Rhijn's paper, and Wilson's calibration of the *p–l* relation, the use of Cepheids to estimate distances was still not freed from suspicion.

In late 1923, Curtis told C. D. Shane at Lick that he doubted whether the cluster variables and the long-period Cepheids were generically related. Adams, Curtis continued, had recently remarked that the cluster variables and long-period Cepheids were 'two different breeds of cats'.[113] Curtis maintained, as he had at the time of the 'Great Debate', that the proper motions of the Cepheids were poorly known, and different authorities gave such different proper motions for the same Cepheids, that the available proper motions could not be relied on to produce a trustworthy calibration. Yet just over 17 months after writing to Shane, Curtis was telling a correspondent that while he had always supported the island universe theory, Hubble's Cepheid observations made the theory's correctness 'doubly certain'.[114] Since the Cepheids now indicated that external galaxies existed, Curtis was prepared to look more favourably on the period–luminosity relationship. Curtis was deeply committed to the island universe theory, and his change of attitude towards the Cepheids was a reflection of this.

The opposition to the Cepheids as reliable distance indicators of some of the advocates of the island universe theory, such as Curtis and Jeans, helps to explain why only Lundmark suggested searching for Cepheids in spiral nebulae. Certainly any astronomer who did accept the Cepheids as distance indicators was likely to have preferred Shapley's model of the Galaxy, and so been inclined against the island universes anyway. Further, the fact that Shapley and Russell were confident of the accuracy of the period–luminosity relation makes intelligible the swift desertion by them of the internal motions. In a choice between van Maanen's measures and the Cepheids they had a much firmer commitment to the Cepheids since much of contemporary astrophysics was intimately linked to the period–luminosity relationship, while the rejection of the measures would not compromise any hypothesis other than that the spirals are not island universes.

That the 100-inch reflector could detect extremely faint Cepheids was known in 1923 since 19th magnitude Cepheids had been detected in the distant globular cluster NGC 7006.[115] But even Hubble was surprised by the Cepheids in spirals, and when in October 1923 he noticed the star in the Andromeda Nebula that proved to be the first Cepheid, he marked it in his notebook as a nova. Only in February 1924, after scanning numerous plates of the nebula, was he so sure of the true nature of his find that he could reveal it to an astronomer outside of Mount Wilson. In 1936 Hubble was to recall that the ending of the debate on island universes was 'an achievement of great telescopes',[116] but Hubble did not claim that without the 100-inch telescope the Cepheids could not have been detected; certainly Hubble was able to observe Cepheids in galaxies with the 60-inch reflector. Yet observing the Cepheids in the Andromeda Nebula was no simple matter, even with the 100-inch. In July 1924 Hubble had secured good plates of the Nebula on 15 successive nights, at the end of which time he had found several Cepheids. However, 'the [variables] are all fainter than [photographic magnitude] 18.0 at maximum and require long exposures in good seeing centred on the outer regions of the nebula'.[117] Also, the careful searches between 1917 and late 1923 for novae on plates of spirals had led to the discovery of a mere three variables.

Could the Cepheids have been found with the aid of the Crossley Reflector at Lick, the bastion of the island universe theory? According to C. D. Shane, a former director of Lick who was himself active in the 1920s, it 'would be a close thing for Hubble's Cepheids to have appeared on Crossley plates of the period'.[118] However, no astronomer at Lick in the early 1920s wanted particularly to photograph spiral nebulae. In 1925, W. H. Wright wrote to Slipher from Lick and told him that since Curtis had left in 1921, no one had 'interested himself especially in the photography of nebulae'. Wright also confided that 'I am technically in charge of the Crossley [reflector], but have no bent for work of that kind'.[119] Thus, while the telescope *was* available to at least try to detect Cepheids, until 1924 the motivation was not present.

Harlow Shapley had no reservations about using the Cepheids as distance indicators and, furthermore, he had been well placed at Mount Wilson to search for them in spirals. As he told Hubble in 1924, during the early years of his Directorship at Harvard he had even collected together several hundred Harvard photographs of the Andromeda

Nebula with various telescopes and magnitude limits:

> A long time ago I had them all assembled for the purpose of examining the region on all to test whether or not any other phenomena like that of 1885 have occurred. The work has not yet been completed, and, moreover, it is probably not very important, for I anticipate a negative result.[120]

In 1969 Shapley remembered that while at Mount Wilson he did not try to find Cepheids in spirals because it was not his assigned job.[121] But his explanation cannot be applied to the period 1921 to 1924 since Shapley was then at Harvard and, anyway, Adams had offered to provide any plates that he wanted. But Shapley was then convinced that the spirals did not contain stars, and so he had no positive reason to search for Cepheids in spirals.[122]

In conclusion, Hubble's success in detecting Cepheids was not simply a result of his having access to the 100-inch telescope: as a follower of the island universe theory, he had been strongly motivated to try and prove the existence of stars in spiral nebulae, and to do so, he had pursued a carefully planned research programme with skill and determination. He had reached his anticipated goal, but by the unexpected route of the Cepheids.

Opposition to the island universe theory

Although after 1925 the island universe theory was accepted by nearly all astronomers, a few sceptics remained. The most persistent of these was van Maanen. When the first Cepheids were found in the Andromeda Nebula van Maanen was not perturbed. As a check on his investigations of the spirals, he had for some time been planning to examine the internal motions of a globular cluster, M 13. In July 1924 he completed these measures and as he had been unable to detect systematic motions within the cluster, it was a 'nice confirmation'. of his measurements of the spirals.[123]

By February, 1925, however, van Maanen was telling Shapley that he was perplexed by the existence of two sets of observations – his own and Hubble's – that led to such radically different conclusions. He had been 'playing again' with his measures of the spiral M 33, but they 'seem to be as consistent as possibly can be'.[124] Van Maanen was now anxiously searching for ways out of the deadlock. First, he wanted to verify if novae in spirals were really comparable with novae in the Galaxy. Secondly, van Maanen described an analysis by Hertzsprung of a star cloud near η Carinae. Here Hertzsprung had discovered some 36 Cepheids, yet

Hertzsprung had suggested that either the majority of these stars did not belong to a single cloud, or the period–luminosity relationship did not hold in this region.[125] Van Maanen wondered if Hertzsprung, the first astronomer to employ the Cepheids as distance indicators, had now shown that the Cepheid criterion was not uniformly applicable throughout the Universe. If so, were Hubble's distance estimates to the Andromeda Nebula and M33 correct? Thirdly, van Maanen told Shapley that he wanted to explore further the possibility that his own measures were subject to a magnitude error. By July 1925 van Maanen had indeed done so. Addressing a Special General Meeting of the Royal Astronomical Society in England, he announced that for three of the seven measured

Fig 16. Adriaan van Maanen (left) and Bertil Lindblad (centre) (Courtesy of A.I.P. Niels Bohr Library, Dorothy Davis Locanthi Collection).

spirals he had found a slight increase in the rotational components of the comparison stars with decreasing magnitude, but the four others showed the reverse effect. He had thereby concluded that the rotational motion, as a whole, could not be explained as a magnitude error. (It is noteworthy that he spoke of rotational motions and *not* stream motions – by now of course Jeans had decided that 'van Maanen's measurements have to go', and the prop that Jeans's theories had provided for the measures had been removed.)[126] This conviction led to a controversy between himself and Hubble, and we shall now follow its course to its public resolution in 1935 with van Maanen's admission that his measures needed to be viewed 'with reserve'.

The dispute between Hubble and van Maanen caused a problem at Mount Wilson that was aggravated by the Observatory's earlier official backing of van Maanen's observations. For example, in the Annual Report of the Observatory for 1923 it was announced:

In summing up the results of this difficult and very important research [of van Maanen's], mention should be made of the strong character of the evidence which leads us to place full confidence in the values obtained.[127]

In consequence of Hubble's research, it now seemed impossible to 'place full confidence' in van Maanen's measures. But what attitude was the Observatory, in particular its Director, to take? At first Adams hoped that a magnitude error might be found which would explain the measures.[128] When this possibility was rejected by van Maanen, Adams merely reported Hubble's and van Maanen's results and did not comment on the controversy.

An extensive investigation of the source of van Maanen's spurious measures has been conducted by R. C. Hart.[129] Hart points out that Walter Baade, an eminent astronomer active at Mount Wilson for many years, later suggested that the measures could have arisen from a number of sources: seeing errors, centring errors and errors produced by Ritchey's method of taking plates of spirals. Baade's favoured explanation was the latter. He believed that the geometrical centres of Ritchey's images did not coincide with the centres of density of the images because Ritchey, in his desire to utilise the best seeing conditions, opened and closed the shutter of his camera a number of times during the course of an exposure. A magnitude error might thus have arisen if sufficient care had not been taken when measuring the plates.[130] But what Baade totally failed to explain was how a magnitude error could have led to motions along the arms of each of the spirals of such remarkable consistency. Also, Ritchey took less than half of the plates van Maanen used for his internal motion studies. Hence Baade's explanation is inadequate. In his own computer aided analysis of the potential error sources, Hart considered the following possibilities:

(1)   Instrumental Sources
 (*a*)   errors due to telescopes
 (*b*)   errors due to plates
 (*c*)   errors due to measuring instruments
(2)   Computation Sources (errors in reduction)
 (*a*)   errors due to invalid procedure
 (*b*)   errors due to neglect of terms in reduction formulae
 (*c*)   errors due to computational blunders
 (*d*)   errors due to poor assumptions
(3)   Personal Sources
 (*a*)   errors due to identification of points on plates
 (*b*)   errors due to choice of objects measured
 (*c*)   errors due to interpretation of results
 (*d*)   errors due to mistakes in position determination
 (*e*)   errors due to bias in measuring

Hart concludes that the internal motions can be explained only as the result of a systematic error which caused van Maanen to add a small component, consistent with the spiral features of the nebula under study, to each point in a spiral that he measured. He argues:

The consistency of the magnitude [of the error] for all the spirals van Maanen measured could be due to the fact that the error need not be especially large – only of the order of the accuracy attainable. The consistency of the directions of the spirals with respect to their arms (exactly the *wrong* direction) could be due to an unconscious bias acquired from merely looking at the plates.[131]

In other words, van Maanen was 'seeing' what he expected to see. There are many similar episodes in the literature of the history of science. For example, S. J. Gould has recently examined a nineteenth century physician's ranking of races by cranial capacity. Samuel George Morton brought together the world's largest pre-Darwinian collection of human skulls and he measured their capacity. Gould discovered that Morton's finding that whites should be placed above Indians, with blacks at the bottom, was in fact the product of apparently unintentional finagling, and the conclusion Morton should have drawn from his data was that all races have approximately equal cranial capacities.[132] Morton had thus unconsciously produced the results to be anticipated in an age when few

Caucasians questioned their innate superiority. In the present chapter, we have already seen that Shapley's periods for the 13 claimed 'cluster variables' in the Small Magellanic Clouds were all badly wrong, yet the stars sat perfectly on the period–luminosity curve. These examples illustrate the enormous influence of prior beliefs and commitments on experimental practice, observation and data-handling, and Gould suggests that unconscious manipulation of data may indeed be common in science.[133] In this light, van Maanen's measures are not so bizarre as one might have assumed, and the hypothesis that van Maanen's measurements were afflicted by an unconscious bias has been strengthened by Stenflo's and Hetherington's analyses of measurements van Maanen made of the shifts of the spectral lines of the Sun produced by the Zeeman effect. Here too van Maanen's measures, which again gave the expected results, seem to have been vitiated by a personal bias.[134] But had his contemporaries publicly advanced such a bias of van Maanen's as the origin of the motions in spirals it would have been tantamount to accusing him of professional incompetence, and the ethics of the astronomical community of the period dictated that before such a claim could be publicly entertained the other possibilities had to be exhausted. Van Maanen's reputation also militated against such a judgement. He was a respected astronomer who was recognised as a most careful investigator of proper motions.

We have seen that Hubble delayed announcing the discovery of Cepheids because they indicated distances to the spirals that were in conflict with those derived from the internal motions. In early 1925 Hubble also held from publication a paper on the external star system NGC 6822 while he searched for an error in van Maanen's measures. Hubble selected for study the measures of M 81 because this spiral had exhibited the largest motions.[135] He knew that van Maanen had made his measures by setting on the centre of density of a star or nebular image, but for images that were well off the telescope's optical axis Hubble said the coma was 'enormous' and the centre of density of an image was at a considerable distance from its optical centre. The apparent luminosity of an object determines the size of its photographic image, and since the comparison stars were, on average, brighter than the measured nebular points, they would be subject to greater distortion; hence the distances between the centre of density of the star images and their optical centres would be different to the corresponding distances for the nebular images. Van Maanen, as we have noted, had considered such magnitude errors but he had argued that they would not affect his results: any systematic difference in the patterns of images on different plates would be symmetrically distributed about the plate centre and so eliminated by the methods of reduction employed. Hubble conceded that if the plates van Maanen had used had been strictly comparable, this would have been so, but the plates were far from strictly comparable. The distortions were not symmetrically arranged about the plate centres, and so Hubble concluded that the differences in magnitudes, and thus image sizes, between the comparison stars and nebular points might thereby have produced a magnitude error.[136] For a time Seares and Hubble tried to persuade van Maanen that such an error was present, but van Maanen was unable to agree with their arguments (as we saw on p. 128).[137]

However, what had begun as a scientific controversy had by early 1925 also taken on a personal aspect and one that led to a very strained relation between the two astronomers, and van Maanen's unwillingness to admit that his measures were spurious has to be seen against this background. In March 1925, in a handwritten appendage to a letter to Shapley, Hubble commented:

Van Maanen takes the idea of a magnitude error in a wildly personal manner. The whole affair is rather a lark, especially as my main purpose is to force him to measure M 13 at the primary focus using bright comparison stars and faint cluster stars. The problem of reconciling the two sets of data is fascinating, but we must be certain of the reliability of the data before we proceed.[138]

Two months later, Shapley himself confided to Seares:

We have just had a good visit with the Hubbles. Hubble's attitude toward van Maanen disturbs me a little, because of my friendship for the latter: Hubble can so well afford to be generous as he has nothing to lose.[139]

In 1926 van Maanen doggedly returned to the possibility that the Cepheids might not be accurate distance indicators. J. Schilt had argued in a paper in the *Astrophysical Journal* that within the Galaxy the *p–l* relation applied only to stars with periods shorter than 10 days: longer-period Cepheids do *not* obey the *p–l* relation; indeed, they are *less* luminous than the shorter-period Cepheids.[140] After reading Schilt's paper van Maanen told Shapley that he intended to test Schilt's findings by determining the proper motions of about 150 galactic Cepheids. Shapley warned van Maanen that Russell and R. E. Wilson had both spoken unfavourably of Schilt's Cepheid studies, but van Maanen replied that he was pressing ahead with the proper motion programme in the free time between his stellar parallax researches.[141] In addition, in 1927 van Maanen again investigated the internal motions of the globular clusters as a check on the possibility of a magnitude error in his measures of the spirals;[142] once more he found no correlation between the magnitude of a star and the size of its rotational component.[143]

As we shall see later, it was not until the early 1930s that Hubble chose to try to resolve publicly the flagrant discrepancy between the implications of his observations of spirals and those of van Maanen's measures of their internal motions. For the present Hubble aimed his efforts at demonstrating the applicability of the principle of the uniformity of nature throughout the visible Universe. By so doing, he was strengthening the reliability of the Cepheids as distance indicators and, incidentally, outflanking and further isolating van Maanen's measures of internal motions.

By the late 1920s Hubble was generally regarded as the leading investigator of extragalactic nebulae (his preferred term for what would now be called galaxies). His prestige and reputation were due in part to his use of the 100-inch telescope at the excellent site of Mount Wilson, and in part to his successes in nebular astronomy, especially his discovery of Cepheids. Hubble's avowed strategy in extragalactic astronomy was to advance in a meticulous step-by-step manner into unknown regions of space; between 1925 and 1929 Hubble did this with four long papers on extragalactic nebulae, and we shall now devote our attention to three of them (the fourth will be considered in the next chapter).

In 1925 Hubble wrote on 'NGC 6822, a remote stellar system', one that he believed was of great significance: 'NGC 6822 lies far outside the limits of the galactic system, even as outlined by the globular clusters', Hubble declared, 'and hence may serve as a stepping-stone for speculations concerning the habitants of space beyond'.[144] Hubble centred his analysis on the 11 Cepheids that he had found in NGC 6822. The period–luminosity relation was 'conspicuously present' and, moreover, the curve he drew for these Cepheids was parallel to that determined by Shapley from the Cepheids in the Galaxy, the globular clusters and the Magellanic Clouds. By employing the period–luminosity relation Hubble estimated the distance of the star system to be 700 000 light years. Furthermore, he claimed that other distance indicators supported this figure: the absolute magnitude of the brightest stars, for example, indicated a similar distance. Hence Hubble concluded:

The principle of the uniformity of nature thus seems to rule undisturbed in this remote region of space. This principle is the fundamental assumption in all extrapolations beyond the limits of known and observable data, and speculations which follow its guide are legitimate until they become self-contradictory. It is therefore a matter of considerable importance that familiar relations are found to be consistent when applied to the first system definitely assigned to the regions outside the galactic system.

Of especial importance is the conclusion that the Cepheid criterion functions normally at this great distance.[145]

In 1926 Hubble's advance into the distant regions of space was carried publicly a step further with his paper on 'A spiral nebula as a stellar system, Messier 33'. Here his methodology was similar to that for his investigation of NGC 6822. First, he analysed the Cepheids that he had detected in M 33. From a study of their light curves and the p–l relation that they exhibited, he maintained that a 'presumption is thus raised in favor of the general validity of the Cepheid criterion which only strong and cumulative evidence to the contrary will destroy'.[146] Hubble again bolstered the Cepheid criterion by examining the bright stars and nebulae within M 33 because these objects offered other criteria for distance estimates. A comparison of the dimensions of the nebulae within the Galaxy and those within M 33, and a comparison of the magnitudes of the brightest stars in the Galaxy and in the spiral, both implied distances to M 33 which corroborated the figure of 850 000 light years derived from the Cepheids.

Three years later, in 1929, the results were published of Hubble's examination of the Andromeda Nebula.[147] To show the reliability of the Cepheids as distance indicators, Hubble here employed the novae that had been observed in M 31, a tactic that had been denied to him in his assault on M 33 because only fragmentary observations of two novae had been at his disposal. A total of 76 novae had been found in M 31 – 63 of them by Hubble – and Hubble affirmed that their absolute magnitudes indicated a distance supporting that derived from Shapley's period–luminosity curve for the Cepheids. He was able to dispose effectively of Schilt's objections to the period–luminosity relation by pointing to collected runs of plates on single nights that clearly exhibited the expected brightness changes of the long-period Cepheids. These studies strengthened the conviction by now held by nearly all astronomers that the spirals are external systems, island universes of gas and stars.

But in the early 1930s van Maanen believed that he had found an 'ally' in his battle against this view. In 1928 Shapley had informed Hubble: 'Curiously enough I find that H. H. Plaskett has grave doubts about the large distances for extra-galactic nebulae. He is sufficiently serious about it that he and Miss Payne are to debate the subject at a colloquium two weeks from today. If he convinces me, I shall cable you!'[148]

H. H. Plaskett, then an Harvard astronomer and later a professor of astronomy at Oxford, continued to oppose the island universes. In 1931 he lectured the Royal Astronomical Society of Canada on 'The nebulae outside the Galaxy'. Plaskett's central thesis was that the advance of astronomy had often been impeded by the uncritical attitude of astronomers, and he illustrated his argument by reference to the extragalactic nebulae. A commentator noted:

After describing Hubble's method of using Cepheid variables, the lecturer pointed out that, possibly on account of the absorption of light in space, the distances so assigned [to the extragalactic nebulae] were too great... The lecturer concluded that the extra-galactic nebulae were probably not island universes, but were rather manifestations of what nature could accomplish with huge masses of gas in those regions of space remote from the disturbing influence of the galactic system.[149]

Plaskett seems to have gained little, if any, support for his views. But van Maanen was still defending his measures and anxiously seeking ways to establish their accuracy. In 1930 he had announced that he had detected the proper motions of the spirals NGC 4051 and M 51.[150] These positive results encouraged van Maanen to measure the proper motions of several more spirals, and in the 1931 Annual Report of the Mount Wilson Observatory Adams commented on these investigations. He even described how van Maanen had measured six points in M 51 which exhibited internal motions in the same direction as van Maanen had found in his original measures of the spiral, and that the mean rotational components of these six points was $0''.015 \pm 0''.003$, as compared with $0''.019 \pm 0''.002$ from the old measures.[151] Adams, the Director of Mount Wilson, was in a predicament. In the very same report, to steer a middle course between van Maanen and Hubble, he praised Hubble's researches on extragalactic nebulae, which of course blatantly contradicted van Maanen's results.

Van Maanen himself told Shapley about his studies of the proper motions of the spirals. Shapley answered: 'I also do not know what to think of your confounded spirals. I am going to mention your preliminary results at the Observatory colloquium on Wednesday, but there is little chance that we can get the universe out of this mess.'[152] In January 1932, van Maanen further disclosed to Shapley that using a pair of plates with a nine year separation he had measured motions in M 33.[153] Once more he had seemingly corroborated his measures of the early 1920s. Despite Shapley's kind words – addressed, we should remember, to a close friend – there seems not the slightest possibility that he was inclined to believe again that van Maanen's measures represented real motions; to have done so would have been tantamount to destroying the foundations of much of galactic and extragalactic astronomy.

Russell also informed Shapley of van Maanen's renewed efforts to bolster his earlier measures. Russell, a frequent visitor to Mount Wilson, was intimately aware of the controversy between Hubble and van Maanen. Van Maanen, Russell told Shapley in 1931, had not found a systematic error in his old measures and he was – presumably as a check on the accuracy of the Cepheids as distance indicators – insisting on a complete survey of all the variables in the Magellanic Clouds. 'Meanwhile', Russell urged, 'it is up to you and Hubble to suggest critical tests of the presence of systematic errors in van Maanen's plates'.[154] While there are no surviving letters between Hubble and Russell on this point, it seems likely that Russell made the same suggestion to Hubble. Certainly Hubble's decision to try and apply the *coup de grâce* to van Maanen's measures dates from about this time. Hubble was also concerned at a slight revival of interest in van Maanen's internal motions,[155] and it is not surprising that by 1932 Hubble was intent upon a frontal attack.

Hubble devoted much effort between 1932 and 1935 to demonstrating the presence and cause of an error in van Maanen's measures. To do this, he and other Mount Wilson astronomers checked four of the seven spirals that van Maanen had analysed in the late 1910s and early 1920s. Hubble lost his conviction that magnitude errors had vitiated van Maanen's measures, nor did he and his colleagues find any signs of internal motions. There were two possibilities: either the internal motions existed and Hubble and his colleagues had made errors which by *extraordinary coincidence* exactly cancelled out the motions to give null results; or they did not exist and van Maanen had made a straightforward error. It seemed methodologically greatly preferable *not* to invoke an extraordinary coincidence; that is, to accept that no internal motions existed. It was indeed a 'masterly manoeuvre',[156] since Hubble had completely cut the ground from under van Maanen. Hubble too wrote a series of manuscripts on the subject of van Maanen's measures, and some of their wording was very sharp.[157] It was, in fact, too sharp for Seares, the editor of the papers written by the Mount Wilson astronomers, who may in consequence even have blocked their publication.

In early 1935 the controversy over van Maanen's measures came to a head. Hubble wanted to publish a strongly worded paper, but Seares and Hale (who still kept a close eye on the Observatory), as is revealed by three letters in the Hale papers, were striving to reach a compromise with him. In the first of these letters – the last two disclose that the institution referred to was Mount Wilson Observatory, the 'two men' were van Maanen and Hubble, and Hubble the 'dissatisfied individual' – Seares explained his problem to Hale:

The following sounds academic; nevertheless, it states, more explicitly than I put it to you yesterday, the general principle that has determined my action in respect to our present troubles.

For two men in the same institution there is opportunity for personal contact and for direct examination of each others results, and hence for the private adjustment of differences in opinion. The institution itself it seems to me, is under obligation to see that all adjustment possible be made in advance of publication. If agreement cannot be attained it may be necessary for the institution to specify how the results shall be presented to the public. In that event, however, there must be opportunity for the expression of individual opinion, but any such expression should be concerned only with the scientific aspects of the question at issue.

The institution has, I think, the right to enforce this procedure; but in certain cases it may be wiser to waive its technical right and say to a dissatisfied individual, 'Print what you like, but print it elsewhere'.[158]

Later the same day Seares told Hale that he doubted whether a settlement could be agreed. 'I am beginning to wonder', Seares again lamented, 'if the solution does not lie in a suggestion to Hubble that he print his investigation elsewhere than in the observatory publications'.[159] The following day Seares despatched to Hale the 'daily instalment' and he proposed that if Hubble desired to publish elsewhere then Hubble should do so.[160]

However, the conflict was publicly resolved a few months later: a compromise was reached with Seares acting as arbitrator between Hubble and van Maanen. An unseemly public exchange was avoided, and in the Mount Wilson Annual Report for 1935, Adams at last criticised van Maanen's measures of internal motions.[161] Furthermore, the 1935 *Astrophysical Journal* contains two brief papers, one by Hubble and the other by van Maanen.[162] The two papers, which were printed adjacently, have been described by Hoskin:

Hubble's, briefly recording the remeasurements and delicately indicating his conclusions; van Maanen's, defensively outlining new measurements in partial confirmation of his earlier results, but conceding that 'it is desirable to view the motions with reserve'.[163]

The public debate over the existence of detectable internal motions in spirals was over, and the last vestiges of public opposition to the island universe theory had been silenced.

Notes

1 H. Jones (1922) 370.
2 Shapley (1919*b*) 266.
3 In recent years a number of researchers have investigated van Maanen's measures and as a consequence there are several very good descriptions of his published papers: see Hart (1973), Berendzen & Hart (1973), Berendzen, Hart & Seeley (1976) and Hetherington (1972). Van Maanen's papers on internal motions between 1920 and 1923 were van Maanen (1921*a, b, c*, 1922*a, b*, 1923*a, b*).
4 van Maanen (1921*b*) 356.
5 van Maanen (1922*b*) 216 and van Maanen (1923*b*) 277–8. Van Maanen's remarks on the various distance indicators to the spirals have been discussed in Hart (1973) 93–6.
6 Hart (1973) 96.
7 van Maanen (1923*b*) 274.
8 van Maanen (1923*b*) 275.
9 *Ibid*.
10 Smart (1924) 334. In 1924 J. Jackson of the Oxford University Observatory concluded that motions of the comparison stars could not explain away the internal motions (Jackson (1924)).
11 van Maanen (1923*b*) 275.
12 Hart (1973) Chapters 4, 8 and 9.
13 van Maanen (1916*a*) 228.
14 These figures were presented in this form in Hart (1973) 71.
15 Jeans (1919) 6.
16 Kostinsky (1917).
17 Schouten (1919*b*).
18 A. van Maanen to H. Shapley, 23 May 1921, Harvard. We shall shortly discuss Lundmark's involvement with the island universe theory.
19 H. Shapley to A. van Maanen, 8 September 1921, Harvard. Maybe Shapley's remark that the measures 'are taken seriously now' implies that he believed they were not taken seriously before.
20 H. Shapley to A. van Maanen, 8 June 1921, Harvard. As van Maanen's measures were recognised to be very important for the island universe debate, one might ask why there was a four year delay between his extensive studies of M 101 and a preliminary check on M 81, and his researches of 1920 on M 33. In 1916 van Maanen had been aware that his researches bore on the island universe theory, and in 1917 he had even attempted to measure the trigonometric parallaxes of the spirals M 51 and M 31. In December of that year van Maanen had reported to Hale that his value for the parallax of the Andromeda Nebula was $\pi = 0''.004 \pm 0''.005$: 'So that we do not know as yet if this is an island universe!' (A. van Maanen to G. Hale, 17 December 1917, Hale). While van Maanen's principal aim in this investigation was to derive a minimum distance for the spirals, he told Hale that he wanted to obtain some plates of the Andromeda Nebula that might be later used for his proper and internal motion studies. Hale replied that 'The parallax of the Andromeda Nebula apparently does not settle the question, and I think you will do well to make plates of the regions showing the large

numbers of condensations [for measurements of proper motions] as you suggest' (G. Hale to A. van Maanen, 26 December 1917, Hale). Yet, even after Hale's prompting, there was still a two year delay before van Maanen measured any more spirals. During this hiatus Adams, second-in-command to Hale, and van Maanen, a Dutchman, held different views on the First World War and this led to acrimony. Van Maanen was not nearly as tough in his attitude towards the Germans as Adams would have liked, and in April 1918 Adams confided to Hale that he sometimes wished van Maanen was not at the Observatory (W. Adams to G. Hale, 18 April 1918, Hale). We will recall from Chapter 2 that in September 1917 Shapley had told Russell that van Maanen harboured doubts about the reality of the motions in M 101. However, in the very same letter, Shapley protested that here was 'another case of a piece of work indefinitely tied up by priority and personality. [van Maanen] has the ambition, skill, and desire to produce further evidence for or against motions, but *he* hasn't the plates. They are in the manger along with their maker' (H. Shapley to H. Russell, 3 September 1917, Harvard). Perhaps Shapley was suggesting that Adams was restricting van Maanen's researches on the spirals by not helping van Maanen to secure suitable plates. The programme of nebular photography at Mount Wilson concentrated on those nebulae whose character was not known, instead of the large, bright spirals of which van Maanen needed photographs (see Pease (1917) and (1920)). Possibly Adams decided not to alter the Mount Wilson programme especially to accommodate van Maanen's research on internal motions. However, it is more likely that, as Professor Owen Gingerich has suggested to me, Shapley was aiming his remarks at Ritchey and not Adams. Certainly van Maanen did not himself take any of the plates of the spirals that he measured. In fact, for the six spirals studied by van Maanen in the early 1920s, the plates were secured by Duncan and Humason. (We should also add that by the early 1920s, when the passions of the War had abated, the relationship between Adams and van Maanen was entirely amicable with van Maanen sometimes visiting Adams's house to play bridge.) Whatever the reason for the delay, by October 1919 Hale was telling Campbell that Mount Wilson was planning an extensive attack on the spirals with special reference to 'internal motion, proper motion, spectra of various regions, novae, etc . . .' (G. Hale to W. Campbell, 8 October 1919, Hale).

21 In 1943, in his short biographical memoir of H. D. Curtis, the then Lick Director R. G. Aitken wrote that for a time only Curtis's colleagues at Lick and a 'few other astronomers' agreed with his views on the island universe theory (Aitken (1943) 280). Perhaps this mistaken belief has its origins in a story Curtis may have told Aitken. In his notes for a series of lectures that he delivered in 1939, Curtis recalled a conversation he had had in 1917 with F. R. Moulton. Moulton had asked him: 'You believe that stars make a Milky Way and that there may be a million other Milky Ways?' Curtis then told his audience that in 1917 'I was almost the only astronomer who held that the spirals are other galaxies' (Curtis, 'Lectures on cosmogony', 1939, Allegheny). Our findings in Chapter 1 show that Curtis's claim is manifestly false: in 1917 many astronomers supported the island universe theory. Further,

in 1924, before hearing of Hubble's Cepheids, Curtis himself wrote that 'many astronomers' held that there was evidence for the existence of external galaxies (Curtis (1924) 3).

22 Haarh & Luplau-Janssen (1922).

23 Öpik (1922) 410. The currently accepted value for the mass of the Galaxy is 2×10^{11} solar masses.

24 Fernie (1970) 1220.

25 Hubble (1936) 87. Öpik was also involved in a controversy with J. H. Reynolds on the inclination of the planes of the spiral nebulae with respect to the Milky Way (Fernie (1970) 1222). Reynolds insisted that the inclinations were not randomly distributed (Reynolds (1920*b*, 1922)), a claim which Öpik vigorously disputed.

26 McLaughlin (1922) 339. In fact, McLaughlin was defending the comparable-galaxy theory.

27 van Maanen (1922*a*) 213.

28 Jeans (1921*a*). See also his address to the Royal Institution in London in 1924 (Jeans (1924)). However, in 1924 he was no longer so confident as he had been in 1919 that the spiral nebulae develop into globular clusters. Now he suggested that the nearer spirals will mingle with the stars of the Galaxy, and that the Galaxy itself may be composed of clusters of stars that have evolved from spiral nebulae. These changes in Jeans's scheme are probably consequences of the need to revise his estimates of the dimensions of the Galaxy. In 1919 he had assumed a diameter of about 4000 pc, and this figure had inclined him to the view that all the spirals lay outside of the Galaxy's boundaries. No one in 1924 could uphold such a small size for the Galaxy and so Jeans was compelled to place the larger and brighter, and presumably nearer, spirals within the Galaxy.

29 Jeans (1923*a*) 72. Smart's check on van Maanen's methods of reduction was stimulated by this paper.

30 Berendzen & Hart (1973) 95. Jeans had spent much of his early career grappling with the new quantum physics, and so was intimately aware of the state of ferment in the physical sciences.

31 Bohr, Kramers & Slater (1924). See also Chapter 4 of Jammer (1966).

32 On Lundmark see Johnson (1961).

33 Lundmark (1920).

34 H. Curtis to K. Lundmark, 30 March 1920, Lund.

35 H. Shapley to K. Lundmark, May 1920, Lund.

36 Lundmark also visited Lowell and Harvard during his stay.

37 Lundmark (1921). While at Lick he took 'slitless' spectrograms of M 33 that convinced him that the star-like points in the nebula were indeed stars. In 1922, in the widely read text-book *Astrophysik*, K. Graff of the Hamburg Observatory calculated a minimum distance to the Andromeda Nebula. He secured his answer of 1 500 000 light years by assuming that the brightest stars in the Nebula were similar to what he said were the brightest stars in the Galaxy, type G giants (Graff (1922) 448).

38 van Maanen (1921*c*).

39 H. Shapley to K. Lundmark, 27 December 1921, Harvard.

40 K. Lundmark to H. Shapley, 2 January 1922, Harvard.

41 K. Lundmark to H. Shapley, 3 January, 1922, Harvard.

42 H. Shapley to K. Lundmark, 10 January 1922, Harvard.

43 Lundmark (1922) 108.
44 H. Shapley to A. van Maanen, 19 June 1922, Harvard.
45 H. Shapley to P. Doig, 21 June 1922, Harvard.
46 H. Shapley to A. van Maanen, 3 July 1922, Harvard.
47 H. Shapley to K. Lundmark, 15 July 1922, Harvard.
48 *Ibid.* Bulletin No. 11 of the National Research Council contains
    Curtis's and Shapley's 'Debate' papers.
49 A. van Maanen to H. Shapley, 21 October 1922, Harvard.
50 H. Shapley to A. van Maanen, 16 May 1923, Harvard. Shapley and
    Lundmark were in fact to become good friends.
51 Lundmark (1927) 17.
52 K. Lundmark to R. Aitken, 31 May 1924, Lick. Lundmark discussed
    his completed measures in Lundmark (1927) 42–6. Although he had
    detected internal motions that followed closely the paths found by van
    Maanen, Lundmark's motions were some ten times smaller than van
    Maanen's and were in his opinion the expression of some unknown
    optical or instrumental effect.
53 Eddington (1917*a*).
54 Eddington (1923) 161.
55 Slipher (1921).
56 In 1922 Slipher and Lampland nevertheless composed the following
    passage for the International Astronomical Union's report on nebulae:
    'The excellent work of van Maanen in the measurement of several
    spirals for internal motions and rotation should be particularly
    mentioned' (*The American Section of the International Astronomical
    Union Report of the Committee on Nebulae*, submitted in March 1922,
    Reynolds papers, R.A.S.). After he had taken charge of the Lowell
    Observatory in 1916 Slipher strove to avoid any sort of controversy,
    and we should read the above passage as a deliberate effort by Slipher,
    who was the Chairman of the American Committee on Nebulae, to be
    as impartial as possible, and not pass his own judgement.
57 A. van Maanen to H. Shapley, 11 August 1921, Harvard.
58 H. Curtis to V. Slipher, 3 June 1924, Allegheny.
59 S. Boothroyd to H. Curtis, 25 February 1924, Allegheny.
60 Shapley (1923*a*) 316.
61 H. Shapley to L. Silberstein, 24 December 1923, Harvard.
62 Reynolds (1923).
63 Lindemann (1923) 354.
64 For example, Shapley told Russell: 'I shall be interested to hear what
    you think of Lindemann's speculations concerning dust clouds. He
    seems to be fairly innocent of the antiquity of these suggestions, and
    the various discussions of the problem during the past few years'
    (H. Shapley to H. Russell, 19 June 1923, Harvard). Russell replied:
    'As regards Lindemann I can't see why he did it. The surface brightness
    of the central parts of spirals is far too great to be accounted for by
    reflected starlight. Seares proved this several years ago, and I
    remember that Lindemann and I discussed this point when I was last in
    London' (H. Russell to H. Shapley, 1 July 1923, Harvard).
65 E. Hubble to H. Shapley, 23 July 1923, Harvard. Writing on August 27,
    Hubble disclosed that a dozen variables had now been located
    (Harvard).
66 Shapley (1923*b*) 2.

67 Shapley (1923*c*). A further indication of the confusion was an apparent contradiction between the internal motions and the spectrographically determined rotations: the spectrographic observations indicated that the spiral arms were trailing the rotation, but van Maanen's internal motions implied that the arms were leading. This was pointed out to Curtis by Joel Stebbins in early 1924 (J. Stebbins to H. Curtis, 22 January 1924, Allegheny), although van Maanen had been aware of it in 1921 (A. van Maanen to H. Shapley, 5 June 1921, Harvard). See also Berendzen & Hart (1973) 74–5.

68 For biographical details on Hubble see Whitrow (1972), Mayall (1954, 1970). In 1976 Donald E. Osterbrock, Director of Lick Observatory, commented that 'Hubble was technically a rather poor observer, as his old Mount Wilson photographic plates show, but he had tremendous drive and creative insight' (Osterbrock (1976) 92).

69 See Seeley (1973) 57–9. In 1922 Russell told Shapley that he had had a number of discussions about nebulae with Hubble: 'Hubble has a fine lot of material and I hope to work with him a good deal. He will go far – he is an able chap' (H. Russell to H. Shapley, 8 February 1922, Harvard).

70 Hubble (1920).

71 Hubble (1922*b*) 421.

72 Hubble (1936) 91.

73 E. Hubble to H. Shapley, 5 July 1922, Harvard.

74 H. Shapley to E. Hubble, 28 July 1922, Harvard. Some forty years later, Walter Baade, himself a leading astronomer of the 1930s, 1940s and 1950s, recalled that 'When I met [Shapley] in Hamburg in 1920, just after his work on globular clusters was finished, I asked him why he had not continued right away with the galaxies. Ritchey's picture of M 33 . . . had just been published, and it showed that the system was simply covered with stars, especially along the spiral arms. When I asked Shapley why he did not study the spiral form of M 33 he told me that the many images were not really stellar, that compared to stars they were 'soft', and quoted Ritchey as authority. Shapley was so much impressed by this that, although I urged him to try it, he was not convinced and made no attempt to study M 33' (Baade (1963) 27). See also Shapley (1919*b*): here he wrote that even in M 33 it was easy to distinguish between the superposed stellar images and the 'softer' nebulous condensations.

75 In 1922 Hubble and Lundmark had indeed proposed a search for novae in several of the brighter spirals for this very reason (*The American Section of the International Astronomical Union Report of the Committee on Nebulae*, p.8, Reynolds Papers, R.A.S.).

76 E. Hubble to H. Shapley, 19 February 1924, Harvard.

77 Note in Hubble's observing book (Hale Observatories).

78 E. Hubble to H. Shapley, 25 August 1924, Harvard.

79 E. Hubble to H. Shapley, 24 September 1924, Harvard.

80 For a discussion of the circumstances surrounding the presentation of the paper see Berendzen & Hoskin (1971). The paper is summarised in *Publications of the Astronomical Society of the Pacific,* **5** (1925), 261–4, and *Popular Astronomy,* **33** (1925), 252–5. The public announcement of Hubble's researches had appeared in the *New York Times* for 23 November 1924, p.6, where it was reported: 'The results are striking in

their confirmation of the view that these spiral nebulae are distant stellar systems.'

81 P. Merrill to H. Curtis, 30 March 1924, Mount Wilson Archives. In November 1924 Merrill returned to this dilemma in another letter to Curtis: 'An interesting program for our weekly Astronomy and Physics Club, on stellar and nebular distances, is being arranged. First, Adams is to outline the methods used for determining astronomical distances, then on subsequent dates Hubble and van Maanen are to follow and present the present status of the problem of spiral nebulae. They will necessarily be on opposite sides of the question, I think, as Hubble has observed quite a number of faint variables in the Androm.neb. that apparently indicate a rather great distance. Van Maanen's measurements constitute about the only evidence for nearness, but they appear to represent a fact of observation, although it seems to be an almost incredible one involving the overthrow of the law of gravitation and goodness knows what else. It is, however, very hard to understand how his results can be obtained unless the apparent motions are real. The chance that it is photographic seems small and that it is instrumental still smaller. I have tried, without success, to get van Maanen to make measures to test the possibility of the results arising from obscure effects in the emulsion, but I must say it is not very likely that such things can cause the apparent motions. Van did go as far as to measure a globular cluster in the same way as he does spirals, and found practically zero rotation with a small probable error, so this fails to support the instrumental or photographic error theory. The thing appears to me to be a real impasse.' (P. Merrill to H. Curtis, 2 November 1924, Allegheny).

82 E. Hubble to H. Russell, 19 February 1925, Princeton.

83 E. Hubble to H. Shapley, 25 August 1924, Harvard.

84 A discussion of, and extensive quotes from, this manuscript forms the first part of a paper by N. Hetherington and the reader is referred to this for a fuller analysis of Hubble's *Notes* (Hetherington (1974 *b*)).

85 Schouten (1919 *b*). Van Maanen was even more dismissive of Schouten's analysis than Hubble. After Lundmark had asked Shapley why van Maanen did not mention Schouten or Kostinsky in his paper on M 51, van Maanen told Shapley that 'Schouten's paper is so short and unimportant, that it might well be ignored. It is also possible that I quite forgot both papers when I wrote my own about M 51' (A. van Maanen to H. Shapley, 18 February 1922, Harvard).

86 H. Shapley to E. Hubble, 27 February 1924, Harvard.

87 H. Shapley to E. Hubble, 5 September 1924, Harvard.

88 H. Shapley to H. Macpherson, 6 April 1925, Harvard.

89 Manuscript of a series of lectures delivered in Toronto now in the Princeton Archives. The quoted remarks are from the 14th lecture delivered in February 1924.

90 H. Russell to E. Hubble, 12 December 1924, Huntington Library.

91 In 1922 Jeans's influence on the Mount Wilson astronomers had been increased still further when he was appointed a Research Associate of the Observatory. Adams had written that 'The opportunity to secure the benefit of his wide knowledge of the problems of cosmogony and stellar dynamics is proving of the greatest value to the members of our scientific staff' (Adams (1923) 184).

92 J. Jeans to H. Russell, 23 October 1924, Princeton. The calculations to which Jeans referred were given in 1925 in a paper in which Jeans admitted that recent results 'published by Hubble and Shapley seem to establish the inaccuracy of estimates I made some time ago of the distances and other quantities associated with the spiral nebulae' (Jeans (1925) 531).
93 J. Jeans to J. Reynolds, 23 October 1924, R.A.S.
94 E. Hubble to J. Reynolds, 17 November 1924, R.A.S. In April 1924 Hubble had visited the United Kingdom and he had met Reynolds at least twice (J. Reynolds to V. Slipher, 15 April 1924, Lowell). Probably Hubble discussed his first Cepheid results with Reynolds then; certainly the tone of the November letter indicates that he did. Hubble had even read a paper on the classification of nebulae to the Royal Astronomical Society on 11 April. His concluding words were that in the present state of knowledge 'speculation on nebulae should not be carried too far' (R.A.S. Meeting, April 11 1924, 143). He did not take the opportunity of announcing publicly his Cepheid results.
95 Reynolds (1924) 147.
96 *Ibid*.
97 Reynolds (1926).
98 Smart (1928) 281.
99 See, for example, Luyten (1926) and Fath (1926) 275–6.
100 Lundmark (1924*a*) 222.
101 Lundmark (1925) 882.
102 Duncan (1922) and Wolf (1923).
103 For example, in 1920 H. C. Plummer, the Astronomer Royal for Ireland and the Director of the Dunsink Observatory, warned Shapley that the 'period–luminosity law is certainly interesting but scarcely so conclusive perhaps as you may think' (H. Plummer to H. Shapley, 3 August 1920, Harvard).
104 In their analysis Kapteyn and van Rhijn termed the short-period Cepheids Class 1 Cepheids, and included in this group those Cepheids with a period between 0.36 and 0.666 days. The Cepheids with a longer period were termed Class 2 Cepheids.
105 Kapteyn & van Rhijn (1922).
106 In 1922 Shapley told Lundmark that 'I am sure that you have not overlooked a point that apparently escaped Kapteyn and van Rhijn – namely, that the peculiar velocities of the short period Cepheids are too high for the application of the method they used. You probably recognize that the derived proper motions indicate high velocity rather than proximity. I have mentioned in print several times these large proper motions and the observed high radial velocities. Recently I have talked the matter over with Hertzsprung, Crommelin, Russell, Hopmann, Baade, Kohlschütter, and others. They all agree that the conclusion of Kapteyn and van Rhijn was not justified. Lindblad wrote me the same' (H. Shapley to K. Lundmark, 15 July 1922, Harvard).
107 Shapley (1922) 70.
108 Fernie (1969) 712.
109 The figures are taken from Gaposchkin & Payne-Gaposchkin (1966). The Gaposchkins' study involved over 560 000 brightness estimates of more than 1500 variables in the Small Magellanic Cloud.
110 This explanation was suggested to me by J. D. Fernie. He writes that

'anyone who has worked in this field knows it is notoriously easy to get the wrong period from limited plate material obtained at the rate of one exposure per night, and I expect Shapley had little trouble in convincing himself that the stars had short periods. No doubt he failed to check on the possibility that longer periods might fit the data better' (J. Fernie to author, private communication). Furthermore, since Shapley believed that the cluster variables were all of essentially the same intrinsic brightness, once he had 'discovered' a variable to be a cluster variable it was not crucial for him to determine an accurate period.

111 Wilson (1923).
112 H. Shapley to P. Doig, 8 May 1923, Harvard.
113 H. Curtis to C. Shane, 10 December 1923, provided by C. D. Shane.
114 H. Curtis to S. Kuftinec, 12 April 1925, Allegheny.
115 Shapley (1923a) 325.
116 Hubble (1936) 83.
117 E. Hubble to H. Shapley, 25 August 1924, Harvard.
118 C. Shane to author, private communication.
119 W. Wright to V. Slipher, 2 June 1925, Lowell. In 1920 Curtis had left Lick to become Director of the Allegheny Observatory.
120 H. Shapley to E. Hubble, 27 February 1924, Harvard.
121 Shapley (1969) 57–8.
122 There is a remarkable story about Shapley and Cepheid variables that is current at the Hale Observatories and known to some other American astronomers. Around 1920, the story goes, Shapley gave Milton Humason some plates of the Andromeda Nebula to examine with the blink comparator. Humason returned these plates to Shapley after he had marked on them some spots that he said were Cepheids. Shapley then carefully rubbed off the marks and explained why they could not be Cepheids.

I have come across no documentary proof from the period to substantiate this story. However, there are some pointers to its possible truth. As we have noted, when Shapley received Hubble's letter announcing the discovery of a Cepheid in M 31 he had replied by detailing the pitfalls of detecting variables in regions of nebulosity. Shapley seems to be politely intimating that Hubble *may* have succumbed to one of these pitfalls (H. Shapley to E. Hubble, 27 February 1924, Harvard). If this is a correct reading of the letter, it lends credibility to the story. Also, in 1920, Humason, who had started his career on Mount Wilson as a mule driver, was relatively untrained in astronomy and Shapley's correction of Humason's 'mistake' would be understandable (in fact Shapley spent much time encouraging Humason's early studies).

Professor Owen Gingerich asked Shapley about this story 'in the early 1970s after his memory was no longer very sharp. Shapley thought about it a few minutes and then said "You know, it just sounds possible", or words to that effect' (O. Gingerich to author, private communication).

123 A. van Maanen to H. Shapley, 23 July 1924, Harvard. See also van Maanen (1925a).
124 A. van Maanen to H. Shapley, 18 February 1925, Harvard.
125 Hertzsprung (1924) 894.

126 van Maanen (1925*b*).
127 Adams (1923) 195.
128 W. Adams to H. Russell, 4 October 1925, Princeton.
129 Hart (1973).
130 Baade (1963) 28–9.
131 Hart (1973) 223.
132 Gould (1978).
133 For a further example of the effect of preconception on observation see
     Gingerich (1975*b*).
134 For descriptions of van Maanen's measurements of the Sun's magnetic
     field see Stenflo (1970) and Hetherington (1975*b*).
135 E. Hubble to H. Shapley, 11 March 1925, Harvard.
136 E. Hubble to H. Shapley, 6 April 1925, Harvard.
137 A. van Maanen to H. Shapley, 20 August 1925, Harvard.
138 E. Hubble to H. Shapley, 11 March 1925, Harvard.
139 H. Shapley to F. Seares, 12 May 1925, Harvard.
140 Schilt (1926).
141 H. Shapley to A. van Maanen, 4 September 1926, Harvard and A. van
     Maanen to H. Shapley, 22 September 1926 and 4 November 1926, both
     Harvard.
142 van Maanen (1927).
143 I.A.U. (1928) 194. Van Maanen announced this result in a letter to the
     1928 International Astronomical Union meeting. It was dated 20
     December 1927.
144 Hubble (1925) 410.
145 Hubble (1925) 432.
146 Hubble (1926*a*) 254.
147 Hubble (1929*a*).
148 H. Shapley to E. Hubble, 10 October 1928, Harvard. Miss Payne was
     later to become Professor C. Payne-Gaposchkin.
149 H. Plaskett (1931).
150 van Maanen (1930*a,b*).
151 Adams (1931) 200.
152 H. Shapley to A. van Maanen, 6 April 1931, Harvard.
153 A. van Maanen to H. Shapley, 9 January 1932, Harvard.
154 H. Russell to H. Shapley, 27 September 1931, Harvard.
155 In 1932, the International Astronomical Union's Commission on
     Nebulae reported that E. W. Brown had extended his previous studies
     'in which he [had] outlined a gravitational system which was devised
     mainly for the apparent internal motions in certain spiral nebulae,
     which had been measured and discussed by van Maanen' (I.A.U.
     (1932) 170).
156 Hoskin (1968) 51.
157 Hetherington has made a thorough examination of these manuscripts
     (now held in the Huntington Library), and his discussion of them is
     contained in Hetherington (1974*b*).
158 F. Seares to G. Hale, 24 January 1935, Hale.
159 F. Seares to G. Hale, 24 January 1935, Hale.
160 F. Seares to G. Hale, 25 January 1935, Hale.
161 Adams (1935) 185. We should note here that R. C. Hart reports that
     two members of the Hale Observatories, H. W. Babcock and
     A. Sandage, told him that Adams suggested to Hubble and van

Maanen that they remeasure the plates of the spirals. Also, they recalled that Adams suppressed the manuscripts prepared by Hubble because they were too inflammatory in tone (Hart (1973) 167). However, the available documentary evidence indicates that Seares advised Adams not to let Hubble publish his manuscripts; and in the 1935 Mount Wilson report Adams noted that Seares had acted as the arbitrator between Hubble and van Maanen.

162 Hubble (1935) and van Maanen (1935).
163 Hoskin (1968) 51.

4

The development of the island
universe theory 1925 to 1931

Once the question of the extragalactic nature of the spiral nebulae had
been effectively settled, astronomers were better able to focus on other
problems posed by the spirals, and in the present chapter we shall be concerned with this shift of problem area, in particular the attempts to classify the spirals. Throughout this book we have seen that the beliefs
astronomers held about the spirals were intimately linked to the current
theories of our own Galaxy. In consequence, the better to understand
the development of ideas about the spirals between 1925 and 1931, we
shall also sketch the changes in galactic theory in these years.

The classification of galaxies
Numerous attempts were made to classify the nebulae during
the nineteenth century and the early years of the twentieth. In the late
1910s Curtis held that nebulae could be divided into three groups: diffuse
nebulae, planetary nebulae and spirals. By the early 1920s, some astronomers, Hubble amongst them, believed that Curtis's classification
scheme was grossly inadequate, and after Hubble in 1924 had all but
ended the debate over the existence of external galaxies, Curtis's scheme
was generally seen as too restrictive. In particular, if an astronomer
wished to trace the course of the evolution of galaxies it would hardly do
to term them all 'spirals' since such a terminology made no attempt to
discriminate between different sorts of galaxies.

Hubble had been interested in classifying the nebulae since the mid-
1910s. The problem was of vital concern to him chiefly because the
nebulae were so numerous that they could not all be studied individually.
It was thus necessary for him to know whether a fair sample could be
obtained from the brighter ones, and, if so, how big such a sample would

have to be.[1] In 1922 he proposed a classification scheme of his own.[2] He divided the nebulae into two main classes according to distribution on the sky: galactic and non-galactic. The non-galactic nebulae were then split into five morphological types: spiral, spindle, ovate, globular and ir-regular.[3] But in 1923 Hubble altered his scheme; now he said he was intent on constructing a classification of non-galactic nebulae analogous to Jeans's evolutionary sequence, but from purely observational material and based on the distinction between amorphous nebulosity and the granular beaded arms of spirals.[4] Jeans's pronouncements on the spirals, as we have seen, were very highly regarded by the Mount Wilson astronomers and it is not surprising that it was to Jeans's theoretical researches that Hubble turned, as had van Maanen, for enlightenment on the course of nebular evolution. Yet when in 1925 a slightly revised version of Hubble's classification scheme was presented to a meeting of the International Astronomical Union, the Committee on Nebulae and

Fig 17. NGC 4649 (centre right) and NGC 4647 photographed in 1920 at Mount Wilson with the 100-inch reflector. Both of these objects at the time were classed as spirals (Courtesy of Mount Wilson and Las Campanas Observatories, Carnegie Institution of Washington).

Clusters rejected it because of its theoretical bias and recommended that a simpler system of a more purely descriptive nature should be used.[5] The scheme as presented was as follows:[6]

A Regular
1 Elliptical   En (n is an index of ellipticity)

2 Spirals
(*a*) Normal Spirals   S
(1) Early   Sa
(2) Intermediate   Sb
(3) Late   Sc
(*b*) Barred Spirals   SB
(1) Early   SBa
(2) Intermediate   SBb
(3) Late   SBc

B Irregular

Despite the criticisms levelled at this scheme, in 1926 Hubble was delighted to secure what he proposed was a demonstration of its usefulness. The empirical justification for the divisions he had adopted was provided, Hubble announced,[7] by the following relationship:

$$m_t = C - 5\log d.$$

Here m_t is the total magnitude of the nebula, d is the diameter of the nebula and C is a different constant for each type of nebula, and this constant changes progressively through the sequence of nebulae.[8] The relationship implied that, as a nebula aged, its diameter and luminosity altered in a well-ordered and predictable manner. Hubble's belief in a smoothly evolving sequence of nebulae was thus reinforced by his discovery and he also emphasised that the relationship indicated a path of development for the nebulae almost identical to that derived by Jeans from his theoretical investigations.

The agreement is very suggestive in view of the wide field covered by the data, and Jeans's theory might have been used both to interpret the observations and to guide research. It should be borne in mind, however, that the basis of the classification is descriptive and entirely independent of any theory.[9]

Indeed, Jeans, having in 1924 abandoned van Maanen's observations, had found in Hubble a new champion. With his enthusiastic adoption of Jeans's theory of nebular evolution, Hubble was also advocating Jeans's interpretation of the elliptical and the central parts of spiral galaxies (which had so far resisted Hubble's attempts to resolve them into stars[10]) as gigantic clouds composed chiefly of gas and dust, and *not* vast collections of stars. For example, discussing the Andromeda Nebula in 1928, Hubble wrote that a 'superficial interpretation' of the data suggested that the nucleus was a star cloud that appeared redder than the spirals' arms because of the relative rarity of blue giant stars.[11] But from the observed shapes of nebulae Jeans had concluded that the nuclear regions of spirals, as well as the elliptical nebulae, must be gaseous, and this explained why they could not be resolved. Hubble wrote:

The novae and spectrum raise difficulties, but these, according to Jeans, are not unsurmountable. The novae, for instance, might possibly be explained as due to the penetration into the gaseous region of stars from the outer portions. This would also account for the lower frequency of novae in highly resolved nebulae (e.g. M 33, NGC 5822 and the Magellanic Clouds), where the unresolved regions are relatively small.[12]

Fig 18. Walter Adams (left) with Hubble (centre) and Jeans at Mount Wilson in 1931 (Courtesy of the Archives, California Institute of Technology).

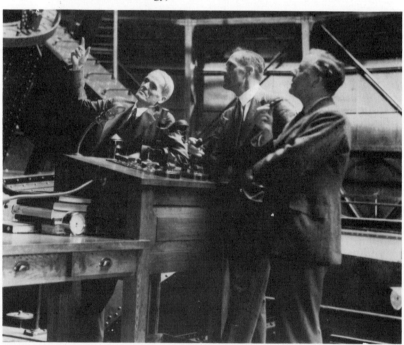

These beliefs help to explain in part why Hubble preferred the term 'extragalactic nebulae', rather than the 'galaxies' championed by Shapley at Harvard (these terms became shibboleths that divided East and West coast astronomers in the United States and it was not until Hubble's death in 1953 that 'galaxies' was universally adopted[13]). Many astronomers nevertheless objected to the hypothesis that the centres of spirals and elliptical nebulae are composed chiefly of genuine nebulosity. In 1927 Bertil Lindblad, a leading European astronomer, termed this the 'great question concerning the nature of nebulae'.[14] While he admitted that Jeans's theory was 'beautiful', his own mathematical analyses of the development of spiral nebulae suggested to him that the apparent nebulosity was in fact constituted of stars of rather smaller intrinsic brightness than the stars in the outer spiral arms.[15] Also, he suggested that extragalactic nebulae evolve from very flattened to less flattened systems; that is, in the direction *opposite* to that predicted by Jeans.

When in 1928 Jeans updated the analyses of his *Problems of cosmogony and stellar dynamics* of 1919 in his somewhat controversial *Astronomy and cosmogony*[16] he underpinned his mathematical researches with Hubble's observations. In *Astronomy and Cosmogony* Jeans again claimed that galaxies could be regarded as 'similar manufactured articles, or as astronomical plants belonging to the same species'.[17] Hubble likewise believed that there is one basic type of nebula and nearly all others are evolutionary variants of this particular type. Taking his lead from Jeans, Hubble argued that an elliptical nebula should be placed at the beginning of the evolutionary sequence, and that it irresistibly expands. The end of the elliptical stage is marked by the build up of arms from the material of the nuclear region, and this process continues until the arms are wide open – the nebula continually increasing in size – and the nucleus has become inconspicuous. The nebula evolves into either a normal spiral or what he described as a 'barred spiral'[18] (although he doubted whether the barred spirals formed a parallel sequence to the normal spirals, being inclined to think that all spirals start with a bar, but that only a few retain them throughout their development[19]). Early in the spiral series the arms begin to break into condensations. These condensations are proto-stars or proto-star clusters, and as the spiral ages the region of star formation creeps inwards.

A classification scheme similar in certain respects to Hubble's was advanced by Lundmark. Indeed, a controversy flared when Hubble accused Lundmark of plagiarising his own scheme after hearing it debated at the 1925 meeting of the International Astronomical Union.

Hubble sent a letter to Lundmark in which, after quoting a footnote he intended to publish in a paper, Hubble angrily exclaimed:

This is a very mild expression of my personal opinion of your conduct and unless you can explain in some unexpected manner, I shall take considerable pleasure in calling constant and emphatic attention, wherever occasion is given, to your curious ideas of ethics. Can you suppose that colleagues will welcome your presence when they realize that it is necessary to publish before they discuss their work?[20]

Lundmark vigorously denied that he had even known of Hubble's revision of the 1922 classification until Hubble had written about it to him.[21] He also insisted that his own classification was not, as Hubble believed it was, 'practically identical' with Hubble's.[22]

Lundmark had arranged the anagalactic nebulae (his preferred term for galaxies) into three groups:
(1)   Globular, elliptical, ovate or lenticular nebulae;
(2)   Spirals or spindles;
(3)   Irregular, chaotic systems which he placed under the heading 'Magellanic Clouds'.
Lundmark's main criterion for classification was the degree of condensation towards the centre of the nebula, whereas Hubble had divided the elliptical nebulae according to their eccentricity, and the spirals depending on their form and the degree of development of the arms.[23] It is significant that Lundmark's scheme was not seen as very similar to Hubble's by contemporaries who pursued nebular problems. Shapley, for example, told Russell:

I am afraid that I have not much sympathy with Hubble's attitude toward Lundmark in this regard. In the first place it is easy to see that the classifications are not at all parallel, and so far as I can see Lundmark's is considerably better than Hubble's ... Hubble believes he can ignore the central condensation of extragalactic nebulae, but no one else does. Lundmark ignores the ellipticity, curiously enough.[24]

In 1927 Shapley himself proposed a new scheme of classification for the galaxies. Shapley attempted to combine what he saw as the virtues of the schemes of Lundmark and Hubble by first dividing the galaxies into ellipticals, spirals and irregulars, and then sub-dividing them according to their magnitude, concentration and ellipticity. But he admitted that his divisions were not suitable for the brighter spirals when they were photographed by powerful reflectors.[25]

Yet each of the classification schemes encountered a major problem: Where in the scheme should the Galaxy be placed? We now turn to some of the attempts to answer this problem.

The anomalous size of the Galaxy

In the late 1910s Shapley had calculated that the Galaxy has a diameter of about 100 000 pc. His distance scale had been founded, as was Hubble's for the extragalactic nebulae, on the period–luminosity relationship of the Cepheids. But Hubble's distances to the extragalactic nebulae implied that they were nearly all far smaller than the Galaxy even if Curtis's modest estimate of 30 000 light years for the diameter of the Galaxy was correct. The following table, given by Hubble in 1926, reveals the size of the discrepancy:[26]

Type	Mean diameter (pc)	Type	Mean diameter (pc)
EO	360	Sa	1450
E1	430	Sb	1900
E2	500	Sc	2500
E3	590		
E4	700	SBa	1280
E5	810	SBb	1320
E6	960	SBc	2250
E7	1130	Irr	1500

Furthermore, Hubble had measured M 33 to have a diameter of 4600 pc and only the Andromeda Nebula, with a diameter of 10 000 pc, began to rival the Galaxy. Hence it seemed that the Galaxy was truly gigantic when compared with its neighbours in space. To Hubble in 1926 this anomaly meant that 'the galactic system must be placed at the end of, if not actually outside, the series of known spirals when arranged according to density'.[27] But by late 1928 he had changed his mind. He asserted that although there was a difference between the dimensions of extragalactic nebulae and the Galaxy, the Andromeda Nebula formed a link between the main sequence of nebulae and the Galaxy. Hubble contended that the Andromeda Nebula – the largest extragalactic system whose distance was known accurately – was actually larger than it appeared on photographic plates: the Nebula's outer regions were too faint to be easily detected, and therefore its dimensions were being underestimated.[28] Hence, while the Galaxy seemed to be some 5 to 6 times bigger in diameter than the Andromeda Nebula, probably the disparity was much less. 'The galactic system must be considered as much larger than M 31', Hubble concluded, 'but the ratio is not greater than that between M 31 and other known extra-galactic systems'.[29]

Shapley, who did not subscribe to Jeans's theory of the evolution of galaxies, responded very differently from Hubble to the anomalous size of the Galaxy. Since 1917 Shapley had believed the Galaxy to be far larger than any other grouping of stars. For example, in 1918 he told a correspondent:

Perhaps we should distinguish 'islands' and 'continents', for in a certain sense I should now consider globular clusters and spirals as small islands in space.[30]

If the spirals are islands, then, according to Shapley, the Galaxy is a continent, a view which Shapley could still maintain even after Hubble's dramatic discovery of Cepheids in the Andromeda Nebula. But in 1930 he revised his old model of the galactic system and proposed the so-called *Super-Galaxy Hypothesis*.[31] The central problem to which he now addressed himself was: is the Galaxy really an 'enormous discoidal star cloud unduplicated by any other visible organization or ... a spiral nebula forty or fifty times the diameter of the average spiral system [?]'[32] Shapley had decided that neither alternative was correct, nor was there any escaping the fact that the Magellanic Clouds and all of the spirals are much smaller than the Galaxy. Instead Shapley hypothesised that the Galaxy is a flattened system of typical galaxies, a super-galaxy. According to this interpretation, several of the distinct Milky Way star clouds are, or had once been, typical galaxies; the local system of stars is itself a galaxy, and the large star cloud in Sagittarius is similar to the Andromeda Nebula. Shapley further claimed there are numerous examples of such clouds of galaxies in the sky.

Shapley reckoned that his new scheme held three great advantages over his old:

(1)   He no longer had to explain the transformation of the globular clusters into galactic clusters.

(2)   The major portion of the nearby obscuring nebulosity is concentrated in the plane of the local system, and if the local system is a galaxy, then this distribution is expected because edge-on spirals possess bands of such obscuring matter around their rims.

(3)   Most importantly, the extraordinary size of the Galaxy was accounted for as one need no longer consider our galactic system to be an organisation of unique size and abnormal consequence.[33]

Eddington's reaction to the anomalous size of the Galaxy was very different to Shapley's, and he dismissed it as a result of insufficient observations. In 1932 he wrote:

The lesson of humility has so often been brought home to us in astronomy that we almost automatically adopt the view that our own galaxy is not specially distinguished – not more important in the scheme of nature than the millions of other island galaxies. But astronomical observation scarcely seems to bear this out. According to the present measurements the spiral nebulae, though bearing a general resemblance to our Milky Way system, are distinctly smaller... Frankly, I do not believe it; it would be too much of a coincidence. I think that this relation of the Milky Way to the other galaxies is a subject on which more light will be thrown by further observational research, and that ultimately we shall find that there are many galaxies of a size equal to and surpassing our own.[34]

Hence Eddington saw no pressing need for the Super-Galaxy Hypothesis. In reply to Eddington's criticisms, Shapley told him:

The chief point in advancing the supergalaxy hypothesis was to place our galactic system in a class with its peers. To be sure, the Coma–Virgo cloud of galaxies is large – millions of light years in diameter – and so are many others; but smaller sizes exist... Our galactic system is of average size, and if the Andromeda Nebula group Messier 33, the Magellanic Clouds, and the globular clusters are also included in our supersystem, in form and population it also compares well with others.[35]

But Shapley's hypothesis failed to win support, and by 1933 he was conceding to a correspondent that although the Milky Way still seemed larger than even the biggest galaxies his own hypothesis of a composite galaxy – which he said was a working hypothesis only – was not very satisfactory. And he thought it equally difficult to expand other galaxies up to dimensions comparable with ours.[36]

Lundmark was also alarmed by the anomalous size of the Galaxy. In the 1920s he had become convinced that some galaxies undergo encounters with their companions in space and that these are important factors in their development. This conviction sprang from his statistical analysis of the galaxies.[37] Another product of his programme was a bold hypothesis: the globular clusters are themselves galaxies.[38] Here was a remarkable return by Lundmark to a position not uncommon a decade earlier, a return he felt compelled to make by the Galaxy's anomalous size. In Lundmark's opinion, the one hundred or so known globular clusters, together with the Magellanic Clouds, the local system of stars and a star system in the direction of Sagittarius, formed a cluster of individual galaxies. Early in 1930 Lundmark discussed his hypothesis with Shapley:

Shapley was not so opposed to the hypothesis as I had expected, but, of course, he believes in 'supra-galaxies' [sic], i.e. he thinks that the Milky Way forms a system of higher order than the ordinary anagalactic objects.[39]

While no doubt Shapley disliked Lundmark's elevation of the globular clusters to the status of galaxies, probably he approved of Lundmark's attack on the view articulated by Jeans and Hubble of the evolutionary uniformity of galaxies. In addition, Lundmark had based his cosmology, as had Shapley, on clusters of galaxies. However, Shapley and Lundmark's hypotheses spawned another anomaly: if the galactic system is really a maelstrom of interacting galaxies, why are the globular clusters arranged symmetrically about this chaotic grouping? Lundmark admitted that he could give no explanation for this 'important fact'.[40]

The Lick astronomer Robert J. Trumpler agreed with Lundmark that the globular clusters are external galaxies; indeed, in 1930 Lundmark told Trumpler that this hypothesis was 'our idea'.[41] In the same year Trumpler argued that the Galaxy is some 10 000 to 12 000 pc in diameter – a size which made it comparable to Hubble's estimates for the Andromeda Nebula (10 000 pc) and M 33 (4600 pc) – and that the globular clusters are extragalactic, forming, together with the Galaxy and the two Magellanic Clouds, a super-cluster. Trumpler, furthermore, had for several years investigated the open clusters of stars within the Galaxy, and he concluded that his results for their space distribution 'support in every way the older views concerning the structure of our Milky Way system and its similarity to a spiral nebula, and they disagree entirely with the more recent conclusions by Shapley and Seares [see p. 159] that the Sun and its surrounding star concentration (local system) are quite a secondary formation in a much larger galactic system over 100 000 parsecs in diameter'.[42] However, most astronomers soon decided that Trumpler's studies of the open clusters served to undermine his own assertions. To understand why this was so, we shall first return to the 1920s and briefly consider the discovery of galactic rotation.

During 1927 and 1928 there was a rapid acceptance by many, probably the majority of, astronomers that the Galaxy rotates differentially. That the galactic system rotates had long been suspected, and its flattened form seemed to be a natural consequence of rotation. For example, in 1914 Eddington had described how the well-known Swedish astronomer C. V. L. Charlier had noticed an effect in the proper motions of the stars 'which might equally well be expressed as a rotation of the stars of the Milky Way in a retrograde direction. Perhaps it would be straining the result too far to regard this as evidencing the truth of our surmise [of the rotation of the Milky Way].'[43] The spectroscopic measures of the rotation of spiral nebulae in the 1910s and 1920s had further assured astronomers that the Galaxy, which many accepted was a spiral nebula, itself rotated.[44]

In 1926, J. H. Oort, a young Dutch astronomer who had trained at Groningen under Kapteyn and who had studied at Yale Observatory from 1922 to 1924, hypothesised that our local system of stars is rotating around the distant centre of the Galaxy.[45] In this way, he was able to explain the widely known, but enigmatic, asymmetry of the motions of the so-called 'high velocity' stars. The stars with velocities exceeding about 65 km s^{-1} were known to display a peculiar asymmetry: when their motions were projected onto the galactic plane, they were seen to be directed almost entirely to one semi-circle of longitude.

Another to study the high velocity stars was Bertil Lindblad. In 1924, Lindblad, an Uppsala trained astronomer who had also spent two years at Lick and Mount Wilson between 1922 and 1924, had been driven to consider a rotation of the Galaxy through his attempts to interpret the phenomenon of star-streaming.[46] By 1925 he had decided that the motions of the constituents of the Galaxy were explicable on the hypothesis that the Galaxy is divided into a series of sub-systems, each of which has rotational symmetry about a common rotational axis. Each sub-system has the same equatorial extent, but possesses a different speed of rotation and hence a different degree of flattening.[47] He reasoned that while high velocity stars do not belong to the same dynamical system as those of low velocity, they must be related to the rest of the Galaxy since their motions are symmetrical with respect to the galactic plane. Lindblad then explained the motions of the so-called high velocity stars by arguing that the Sun and other low velocity stars in fact have high speeds of rotation about a remote centre, and that the so-called high velocity stars have much smaller speeds of rotation (and so form a more nearly spherical system), thereby falling behind as the Sun overtakes them. The high velocity stars will thus appear to move asymmetrically. Further, Lindblad (as Oort was soon to do) calculated the dynamical centre of the Galaxy to be very close in galactic longitude to the centre of the globular cluster system as determined by Shapley.[48]

Oort was deeply influenced by Lindblad's researches, and in 1927 he announced that through his attempt to verify directly Lindblad's theory of galactic dynamics, he had secured firm evidence of a differential rotation of the Galaxy.[49] Oort had found that the proper and the radial motions of fairly distant stars exhibited the small but systematic effects to be expected of differential rotation. Moreover, the validity of Oort's

inferences about galactic rotation was corroborated by other astronomers, particularly by the Canadian J. S. Plaskett who analysed the radial
motions of hundreds of O- and B-type stars in terms of the Oort–
Lindblad theory.[50] The detection of galactic rotation had been very much
'in the air' (although some of the views were substantially different from
the theory advanced by Oort and Lindblad) and Oort had presented
what was generally seen as its observational proof. In addition, Oort and
Lindblad had seemingly shown that the Galaxy rotates about a point that
lay in almost the same direction as Shapley's proposed direction to the
galactic centre. Nevertheless, there was one point where the researches
of Oort and Lindblad did not mesh with Shapley's model: the distance of
the Sun from the centre of the Galaxy. Oort had reckoned that the
distance to the centre about which the stars rotated was roughly 6000 pc,
about one-third of the size of Shapley's estimate.[51] But despite this discrepancy, the discovery of differential galactic rotation swept away the
opposition to the eccentric position of the Sun within the Galaxy.

In 1930 the Halley Lecture at Oxford was delivered by Eddington who
took as his subject 'The rotation of the Galaxy'.[52] Here he confidently
advocated the theory that the Galaxy rotates differentially. Eddington
also claimed that the values that Oort and Shapley had deduced for the
distance to the centre of the Galaxy agreed within the errors of their
independent methods. Shapley and Oort's estimates were, furthermore,
soon to be brought into a close agreement by the demonstration of the
existence of a general interstellar absorption.[53]

In the late 1910s Shapley had found no evidence of significant absorption in his examinations of the globular clusters, and he had proceeded to
argue that except for isolated dark clouds, space is effectively trans-
parent. This view continued to be very influential until the implications
of R. J. Trumpler's study of the open clusters within the Galaxy had been
fully grasped.[54]

As his chief working hypothesis Trumpler had taken the open clusters
of similar constitution to have, on average, the same dimensions; by
comparing the observed angular diameter of a cluster with the average
linear diameter of the sub-class to which the cluster belonged, he derived
a value for the distance to the cluster. Trumpler had also examined the
magnitudes and spectral types of the stars within the clusters. Then by
constructing a diagram of the magnitudes of a cluster's stars against their
spectral types – a Hertzsprung–Russell diagram[55] – and comparing the
observed diagram with a standard diagram, Trumpler secured another

value for the cluster's distance. He thereby found that the two distance indicators gave systematically different answers: the more distant the cluster, the more the two distance values differed. Trumpler argued (and astronomers soon agreed) that the reason for this deviation was that interstellar absorption diminishes the brightness of the cluster stars but does not appreciably affect the measured diameters of the clusters.[56] Absorption, then, made the distance to a cluster obtained from its Hertzsprung–Russell diagram appear greater than it actually was. Also, Trumpler contended that the well-known discrepancy in open clusters between the observed colour indices of stars and their colour excesses could readily be explained as a consequence of interstellar absorption.

Although, with the benefit of hindsight, a number of earlier investigations can be seen to have pointed towards the existence of a general interstellar absorption, it was Trumpler's analysis of the open clusters – probably because it was more extensive and complete than earlier researches – that convinced astronomers of the presence of obscuring matter throughout the Galaxy. We should emphasise that although the existence of localised dark absorbing clouds had been admitted for many years, until Trumpler's study the reality of an appreciable general absorption had been widely denied.[57]

In his calculation of the size of the galactic system Shapley had not allowed for the dimming effect of a general interstellar absorption[58] and as a result he had overestimated the distance of the Sun from the centre of the Galaxy. Yet the 'discoveries' of galactic rotation and interstellar absorption were instrumental in bringing into wide acceptance the two central features of Shapley's Big Galaxy – the eccentric position of the Sun and the role of the globular clusters in outlining the Galaxy.

This shift had also been aided by Seares's studies of galactic structure. In 1927 Seares had used the tools of statistical astronomy to explore 'Some structural features of the galactic system'.[59] He had found that although there was a local maximum to the star density, the contours of equal star density that he had traced indicated that the centre of the galactic system lay at a far greater distance than the local centre. Although Seares stressed that he could not fix the actual distance to the centre, he concluded that his star counts implied that it lay in the direction Shapley had deduced for the centre of the Galaxy. Here was powerful evidence in favour of the Big Galaxy and the eccentric position of the Sun, evidence secured with the methods of statistical astronomy whose results had previously seemed to be in flagrant disagreement with

Shapley's model. Certainly this analysis pleased Shapley who told Seares that he had read 'with much interest your deduction of the direction to the center of the galactic system, and of its accordance with the rotational values by Lindblad and Oort'.[60]

In the early 1930s there was, nevertheless, still some dispute about the size of the Galaxy. What many continued to find hard to accept about Shapley's estimate was why the star clouds should extend out even to an appreciable fraction of the distance of the remotest globular clusters.[61] An answer to this query was still not generally agreed by 1935 when J. S. Plaskett noted:

While Shapley appears to consider [the Galaxy] to extend nearly as far as the most distant clusters, more conservative opinion might look upon the outermost members as stragglers, and limit the boundary to the position where the clusters begin to condense.[62]

Plaskett also reported that astronomers usually took the distance to the galactic centre to be about 10 000 pc and that nearly all would agree that the Galaxy's diameter was not greater than 40 000 pc while some would estimate it at 30 000 pc. In fact, it was not until 1952 and Walter Baade's discovery that there were two classes of long-period Cepheids (due to their belonging to two different types of stellar population) that the size of the Galaxy was brought into line with the size of similar types of other galaxies. Baade had found that two Cepheids of the same period but of different population type possessed markedly different luminosities, and in consequence the distances to external galaxies had been underestimated, and thus their dimensions needed to be increased.

During the early 1930s Baade's studies lay in the future, and despite the uncertainty about the size of the Galaxy, it seemed beyond doubt that if contemporary observations were to be trusted the Galaxy was larger than any other galaxy. Thus the island universe theory had now become very different from its predecessor of the late 1910s. Then most astronomers had equated the theory with the comparable-galaxy theory, but by the early 1930s many thought such an identification impossible. The supposed islands were really islands when compared with the continent of the Galaxy. Moreover, Hubble and Jeans and their followers believed the elliptical galaxies and the centres of the spiral galaxies to be composed chiefly of gas and dust and not stars. And so by about 1930 the island universe theory seemed emasculated when compared with its forerunner of the late 1910s. Although one may correctly claim that in the early 1930s it was accepted by almost all astronomers, during the late

1910s and 1920s the content and interpretation of the theory had substantially changed.

Notes

1 Hubble (1936) 36.

2 Hubble seems never to have accepted Curtis's classification, and in his dissertation he had employed Wolf's elaborate scheme which identified 23 basic types of nebula: Hubble (1920).

3 Hubble (1922*a*). See Berendzen & Hart (1971) for an account of Hubble's classification schemes and some of the other schemes proposed in the period.

4 E. Hubble to V. Slipher, 4 April 1923, Lowell. The distinction between the 'amorphous' and 'granular' arms was explained by Jeans as two different stages in the formation of stars in the arms. In the amorphous arms stars had not started to form, but in the granular arms stars had begun to condense out of the nebulous material.

5 I.A.U. (1925) 206.

6 Berendzen and Hart point out that the only changes in his late 1923 classification – the scheme he presented to the I.A.U. in 1925 – and his 1926 classification were that 'n was allowed to take the integral values between 0 and 7 instead of between 0 and 8 in the 1923 manuscript, and a few of the examples were changed' (Berendzen & Hart (1971) 118).

7 Hubble (1926*b*) 346. Hubble excitedly announced his find to Slipher and an excerpt from this letter is printed in Berendzen & Hart (1971) 114.

8 In what follows, 'nebula' implies 'extragalactic nebula' unless otherwise indicated.

9 Hubble (1926*b*) 324.

10 The possible exception to this was M 87: see Hubble (1923).

11 Hubble (1929*a*) 149.

12 *Ibid*. By 1936 he was writing that his inability to resolve the ellipticals did not necessarily mean that stars are absent: 'it merely indicates that, if stars are present, the brightest of them are fainter than the brightest in the later spirals'. In fact, it seems he now favoured this hypothesis despite labelling it 'speculative' (Hubble (1936) 54).

13 The dispute over the use of 'galaxies' or 'extragalactic nebulae' is merely one manifestation of the rivalry, sometimes intense indeed, between Shapley and Hubble that began in the late 1920s and persisted throughout Hubble's life. But in 1928 Hubble told Shapley that the term 'extra-galactic nebula' was unsatisfactory, and that he was inclined to restrict the use of 'nebula' to imply the galactic variety. He thought Shapley's suggestion of 'galaxy' was rather awkward, 'but so far appears to be the least objectionable', and that unless something better turned up he was inclined to adopt it (E. Hubble to H. Shapley, 2 November 1928, Harvard). By May 1929 Shapley had plumped for 'galaxy' in order to avoid using 'nebula' and 'universe' (H. Shapley to E. Hubble, 29 May 1929, Harvard).

14 Lindblad (1927) 421. On Lindblad see Oort (1966).

15 He also argued that the arms wind out from the centre of a spiral in the same direction as the rotation of the spiral. That this was the same direction that van Maanen had found from his investigation of internal

motions did not lead to a revival of support for van Maanen's measures.

16 Jeans's biographer, the brilliant mathematician and theoretical astronomer E. A. Milne, remarked some 24 years after its publication that 'it is not written with the sure physical grasp that we should have expected from one of Jeans's rank' (Milne (1952) 57).

17 Jeans (1928) 19. By now Jeans was speculating that at the very centre of extragalactic nebulae sat gigantic white dwarf stars. Indeed, he went so far as to write that 'apart from their rotation and outer structure, the extragalactic nebulae may be regarded as being white dwarfs of colossal mass' (Jeans (1928) 356). This speculation gained little, if any, support.

18 We are here following Hubble's terminology. Curtis had earlier described these as φ type spirals.

19 Hubble (1926*b*) 353.

20 E. Hubble to K. Lundmark, 26 August 1926, Lund. In the following year Hubble complained to Shapley that none of Lundmark's studies could be trusted since in all of them he mixed 'facts with fancies' (E. Hubble to H. Shapley, 29 April 1927, Harvard).

21 Lundmark (1927) 24.

22 E. Hubble to V. Slipher, 22 June 1926, Lowell. Hubble used the same expression in a letter to Reynolds (E. Hubble to J. Reynolds, 22 June 1926, R.A.S.).

23 Lundmark (1926*a*). He was also of the opinion that the ellipticals had higher radial velocities than the spirals, which in turn had larger radial velocities than the Magellanic Cloud types (Lundmark (1926*b*) 2).

24 H. Shapley to H. Russell, 15 August 1927, Harvard.

25 Shapley (1927). In this paper he expanded the views he had expressed to Russell (see note 24 above). Shapley had designed his scheme for the photographs of nebulae taken with the 24-inch Bruce refractor of the Harvard Observatory.

26 Hubble (1926*b*) 362. At the same time, Lundmark was arguing that ellipticals, on the average, are 10 times smaller than the spirals: Lundmark (1926*b*).

27 Hubble (1926*b*) 362.

28 Hubble (1929*a*) 156.

29 Hubble (1929*a*) 157.

30 H. Shapley to H. Macpherson, 10 December 1918, Harvard.

31 Shapley (1930*c*).

32 Shapley (1930*a*) 212.

33 Shapley (1930*b*) 144.

34 Eddington (1933) 4.

35 H. Shapley to A. Eddington, 14 April 1931, Harvard.

36 H. Shapley to H. Plaskett, 18 July 1933, Harvard.

37 See, for example, Lundmark (1926*b*, 1927).

38 Lundmark (1930).

39 K. Lundmark to R. Trumpler, 6 January 1930, Lick.

40 Lundmark (1930) 30.

41 *Op. cit.* (note 39).

42 Trumpler (1930) 185.

43 Eddington (1914) 216. A later example is Fotheringham (1926).

44 See, for example, Campbell (1926) 82.

45 Oort (1926) 63–6. On Oort's studies see Berendzen, Hart & Seeley (1976), section 2, Chapter 2.

46 See, for example, Lindblad (1925*a*). An early product of Kapteyn's scrutiny of the proper motions of the stars had been his discovery, announced in 1904, that the stars tended to move in two distinct and diametrically opposed directions. Kapteyn supposed this phenomenon to be the result of two once distant but now intermingled streams of stars moving relative to one another: see Kapteyn (1913). While an explanation of Kapteyn's find became one of the outstanding problems of galactic astronomy, the immediate and surprising message of the star streams was that the stars do not move randomly, as had been assumed, but that there is an order amidst the apparent jumble of stellar proper motions.
47 Lindblad (1925*b*).
48 There was, however, at first a basic difference between Oort's theory and that of Lindblad. As Oort recalled in 1971: 'Lindblad had tentatively supposed that the principal part of the greater galactic system is formed by an ellipsoid of constant density, and that as a consequence its rotation is approximately the same as that of a solid body. The systematic effects I found in 1927 indicated that the actual motions were quite different, and that the angular velocity increased strongly toward the center of the Galaxy, indicating a large concentration of mass in the central area' (Oort (1972) 261).
49 Oort (1927*a*) 276.
50 J. Plaskett (1928).
51 Oort (1927*b*) 88.
52 Eddington (1930). However, he did strike one note of caution when he warned that 'a thread of insecurity runs through the whole fabric' due to the lack of evidence from the southern hemisphere (Eddington (1930) 18).
53 It is worth noting that the 'failure' of the classical methods of statistical astronomy to model accurately the Galaxy was due principally to a neglect of interstellar absorption and the belief that the Galaxy could be described at least in a first approximation by two simple functions D (r, b) and ϕ (M) (see also Chapter 2).
54 J. Plaskett (1935) 11 and Seeley (1973) 127–32. J. S. Plaskett's Halley Lecture of 1935 was described by Hubble as 'a simple, clear, and authoritative statement of the present conception of the galactic system' (Hubble (1936) 129).
55 At this time the term in use was 'Russell diagram', only later was this replaced by 'Hertzsprung–Russell diagram'.
56 Trumpler (1930) 163–7.
57 For an excellent discussion of the development of studies of interstellar absorption see Seeley (1973).
58 In 1929 Shapley was still arguing for the effective transparency of space, although he did accept that obscuring matter hides the centre of the Galaxy: see Shapley (1929*a*) and Shapley & Ames (1929).
59 Seares (1927).
60 H. Shapley to F. Seares, 3 November 1927, Harvard. Seares did not accept, as Shapley did, that the Galaxy is an amalgam of clusters. Rather, to Seares it was a highly resolved spiral resembling M 33 and with a diameter of 60 000 to 90 000 pc. Indeed, he judged that the trend of cosmological thought was towards 'parallelism of structure in the stellar system and in spiral nebulae' (Seares (1927) 171). Shapley could

not accept a spiral pattern for the Galaxy and told Seares that 'a nasty mixture of star clouds and confusion has appealed to me more than a regular spiral' (H. Shapley to F. Seares, 3 November 1927, Harvard).

61 See Campbell (1926) 85.
62 J. Plaskett (1935) 23.

5

The velocity–distance relation:
its origins and development to 1931

We discussed in the last chapter the changes in the island universe theory between 1925 and 1931, but we did not examine what has proved to be the most far-reaching element in these developments; the establishment of a linear relation between a galaxy's distance and the redshift of its spectral lines. We shall focus on the observational investigations that led to the relation. Hence, some mathematical models that were later realised to be of great importance for theoretical studies (such as Friedmann's expanding models of the early 1920s), but that have been judged to have had little influence on the observations that were made in the 1920s and early 1930s, will not be treated in detail.

The velocity–distance relation has had an enormous influence on cosmology. Although it has since been presented in a variety of forms, this remarkable relation has been central to all cosmological studies since the late 1920s. It has been said that 'the existence of a linear relation between redshift and distance is the most spectacular discovery in astronomy that has been made during the past 60 years',[1] and the eminent cosmologist G. J. Whitrow writes that this relation has come to be generally regarded as the outstanding discovery in twentieth century astronomy:

It made as great a change in man's conception of the universe as the Copernican revolution 400 years before. For, instead of an overall static picture of the cosmos, it seemed that the universe must be regarded as expanding, the rate of the mutual recession of its parts increasing with their relative distance.[2]

Radial velocities of spiral nebulae
As we will recall from Chapter 1, the first spectroscopic observation of the radial velocity of a star was made in 1868 by William Huggins. The discovery of the radial velocity of a spiral proved much

more difficult, and the first successful measurement was not made until 1913 when V. M. Slipher startled astonomers with his finding that the Andromeda Nebula was moving towards the Earth at 300 km s⁻¹.

By 1915 a few astronomers had decided that sufficient radial velocities had been measured to permit a determination of the Sun's motion with respect to the system of spiral nebulae. The spirals were by this time widely suspected to be island universes and so these determinations possessed cosmological overtones: if the solar motion deduced from the spirals differed from that derived from stellar motions, the implication would be that the spirals formed a system independent of the stars. The method of analysis was straightforward and then very well known to astronomers. If it is assumed that the observed radial velocities of the spirals are due to their own random motions and the drift velocity of the Sun, then all that is required to secure the solar motion relative to the spiral nebulae is to solve a set of equations of the form

$$V = X \cos \alpha \cos \delta + Y \sin \alpha \cos \delta + Z \sin \delta$$

(V is the observed radial velocity of a spiral, and α and δ are its right ascension and declination; $-X$, $-Y$, and $-Z$ are the equatorial components of the solar motion with respect to that spiral).[3]

Although two papers appeared in 1916 that included such computations, one by Truman[4] of the Iowa State Observatory in the United States and the other by Harper and Young[5] of the Dominion Observatory in Canada, they were probably predated by those made by Hertzsprung. The results of Young and Harper, Hertzsprung confided to Eddington, were of little value; he had performed such a determination 'some time ago' from the radial velocities of 13 spirals, but the uncertainties involved in the calculation deprived his answers, as well as those of Young and Harper, of any accuracy, and he had not wanted to make them public.[6] Truman, Young and Harper, however, believed their results underpinned the hypothesis that the Galaxy is drifting through the system of randomly moving spirals since they had found its drift velocity – assuming that the Galaxy's drift velocity is little different from the Sun's – to be similar to the radial velocities of the spirals.

In 1916 Slipher protested to Campbell that these calculations were premature and that the data were 'too scanty and particularly its distribution too poor to give anything of value for the magnitude and direction of the drift [of the Galaxy]'.[7] Yet only a few months later Slipher himself performed the very same calculation. Although he had at his disposal the radial velocities of 25 spirals, his values for the speed and direction of the solar motion were similar to the previous determinations.[8]

As Slipher doggedly and skilfully observed further spirals, a curious pattern began to emerge: all but a few were receding. In 1914 Slipher had presented a list of the radial velocities he had measured[9] (a negative value indicates a velocity of approach, and a positive one a velocity of recession):

NGC	Velocity (km s⁻¹)
221	−300
224	−300
598	—
1023	+200 roughly
1068	+1100
7331	+300 roughly
3031	+ small
3115	+400 roughly
3627	+500
4565	+1000
4594	+1100
4736	+200 roughly
4826	+ small
5194	± small
5866	+600

 ←spirals

The radial velocities of the spirals were also generally very much greater than those of the stars or gaseous nebulae. In fact, the astonishingly large sizes of some of the spectral shifts prompted some astronomers to query the Doppler, or velocity, interpretation of them.

One such was the Lick astronomer G. F. Paddock. In 1916 Paddock computed the solar motion from the radial velocities of the spirals by solving equations of the form

$$V = X \cos \alpha \cos \delta + Y \sin \alpha \cos \delta + Z \sin \delta + K.$$

In the process he employed what proved to be a crucial innovation: the K term. Here Paddock was actually exploiting an established and widely accepted technique. In 1911, Campbell, Paddock's Director at Lick, had introduced the K term to remove an apparently constant redshift (equivalent to a Doppler shift of 4 km s⁻¹) in the spectral shifts of type B stars.[10] Campbell contended that the constant K term (possibly due to peculiar pressure effects in their atmospheres) had to be extracted from the B stars' spectral shifts before he could use them to calculate the solar motion. Paddock likewise used the constant K term to remove what he saw as a non-random recessional component from each spiral radial velocity, but he had little faith in the reality of this component. He expected that as more spirals were observed the average value of the radial velocities would decrease and the K term disappear: its necessary presence in his own calculations was merely a manifestation of an insufficient number of observations.[11]

After Paddock's paper the K term was generally admitted to be a component of a spiral nebula's redshift and it was included in some form in all subsequent attempts to determine the motion of the Sun from the radial velocities of the spirals.

A contribution similar in many respects to Paddock's was C. A. Wirtz's paper of 1918 'Über die Bewegungen der Neblflecke'.[12] Wirtz too used a constant K term. He, like Paddock, did not commit himself to interpreting it as a consequence of a systematic recession of the spirals from the Galaxy. But Wirtz's papers on the motions of the spirals made little impact on the leading students of the spiral nebulae. Maybe this was because he had also derived the solar motion using the very poorly known proper motions of the spirals without emphasising the uncertainty of his data.[13] Then, the following year, Lundmark too computed the solar motion with respect to the spirals. Although he used three more radial velocities than Wirtz had, he arrived at almost the same answers.[14] In 1921, Wirtz reworked his own calculation with 29 radial velocities, but his values for the K term and the solar motion differed little from his previous results.[15]

By the early 1920s, then, a method for calculating the solar motion from the radial velocities of the spirals had been generally agreed upon. An important change within this computational scheme was to arise from attempts to correlate the radial velocities with other parameters, in particular the distances of the spirals. To place this development into context we shall have to analyse the theories that helped to guide it.

Cosmology and general relativity

In 1917 Albert Einstein had written on 'Kosmologische Betrachtungen zur allegemeinen Relativitätstheorien'.[16] As W. H. McCrea has pointed out, Einstein intended his paper to be chiefly a record of the thinking by which he had discovered much of the true nature of general relativity.[17] Here Einstein also explored briefly the implications for cosmological investigations of the general theory of relativity. He decided that the general theory, over which he had toiled for the previous decade, permitted him to speculate on the very size and nature of the physical universe. By making this daring commitment he effectively founded modern mathematical cosmology.

In 1915 Einstein had been in possession of field equations (which he had shaped to agree with the Newtonian law of gravitation for weak fields) of the form

$$R_{mn} - \tfrac{1}{2}g_{mn} R = - \kappa T_{mn}.$$

In these equations, which amount to ten equations for the ten unknowns g_{mn}, κ is a constant; T_{mn} is the energy–momentum tensor; g_{mn} is the metric tensor (which specifies the gravitational field and the scale of time and space intervals); and R_{mn} is the Ricci tensor which is obtained from the Riemann–Christoffell tensor R^p_{qrs} (a tensor constructed solely from the components of the metric tensor and its first and second derivatives with respect to the co-ordinates). By 1917, however, he had altered the equations to

$$R_{mn} - \tfrac{1}{2} g_{mn} R - \Lambda g_{mn} = - \kappa T_{mn}$$

where Λ is the so-called cosmological constant. As the basis of the model of the Universe that he expounded in his 1917 paper, Einstein assumed a uniform distribution of matter in static equilibrium. Einstein's motive for considering such a model was that he was deeply concerned about the boundary conditions he should impose on his field equations. Einstein argued in his general theory that as a consequence of the gravitational fields in the Universe the space–time continuum was 'curved'.[18] Although he wrote that the curvature is variable in time and space, he claimed that he could roughly approximate the actual curvature by means of a spherical space. He thus decided not to *solve* the boundary value problem, but rather to *dissolve* it 'by regarding the Universe as a continuum closed with respect to its spatial dimensions'.[19] This was Einstein's dramatic starting point, and he believed the most significant aspect of his paper. The size he then determined from his solution to the field equations for the spatial extent of the Universe hardly mattered. As he explained in 1921, his calculations indicated that the Universe has a diameter of roughly 100 million light years. This followed from:

mathematical calculations which I presented in 'Cosmological considerations arising from the general theory of relativity', in which the figure I have just quoted is not given. The exact figure is a minor question. What is important is to recognize that the universe may be regarded as a closed continuum as far as distance measurements are concerned.[20]

Here was an immense conceptual shift: a move from an infinite to a bounded universe. During the next decade a number of astronomers and mathematicians felt compelled to follow the path taken by Einstein and accommodate this shift within their thinking.[21]

Einstein was little concerned in his 'Kosmologische Betrachtungen' with the question of whether his model was compatible with contemporary astronomical knowledge. He did, nevertheless, discuss his reasons for introducing the Λ term. It is not, he stressed,

> justified by our actual knowledge of gravitation. It is to be emphasised, however, that a positive curvature of space is given by our results, even if the supplementary term is not introduced. That term is necessary only for the purpose of making possible a quasi-static distribution of matter, as required by the fact of the small velocities of the stars.[22]

The effect of the Λ term was thus to produce a repulsive field to oppose the gravitational field; without its presence Einstein calculated that the stars could not remain in equilibrium, to him an unacceptable conclusion. The size of the Λ term, moreover, defined the mean density of matter as well as the volume of the Universe.

Einstein quickly seems to have doubted his decision to employ the Λ term. Certainly he felt that it impaired the simplicity and elegance he believed all fundamental physical equations should possess, and in 1919 he remarked that he hoped soon to expunge it from his field equations.[23] Despite Einstein's misgivings, the Λ term became a major feature of mathematical cosmology during the late 1910s and 1920s, a cosmology whose foundation-stone, Einstein's 1917 paper, was fashioned in ignorance of the latest astronomical data and designed in accordance with Einstein's assumption of a static universe.

De Sitter's solution

Some of the most important sources of inspiration for Einstein in his construction of the general theory of relativity were the writings of the philosopher and physicist Ernst Mach. Mach had argued that inertial forces are caused by distant masses in the Universe[24] and Einstein believed firmly that he had produced, in accordance with what he called 'Mach's Principle',[25] a theory of gravity that functionally related the inertial properties of space–time to the material distribution of the Universe. However, in 1917 the Dutch astronomer Willem de Sitter secured another solution to Einstein's field equations that blatantly contradicted Mach's Principle.

De Sitter was a Foreign Associate of the Royal Astronomical Society when in 1916 he received a copy of one of Einstein's papers on general relativity. De Sitter sent details to the Royal Astronomical Society's Secretary, Eddington, who then asked him to write on the general theory for the *Monthly Notices of the Royal Astronomical Society.*[26] In consequence, during 1916 and 1917 de Sitter composed three long papers on the general theory. Einstein and de Sitter corresponded in these years, and before de Sitter wrote the second paper they had several long conversations at Leiden in Holland.[27] In the third paper de Sitter explained

Fig 19. Einstein with de Sitter and Eddington. From left to right, Einstein, Eddington, Paul Ehrenfest, H. A. Lorentz and de Sitter (Courtesy of A. I. P. Niels Bohr Library).

his own solution to the field equations: the empty universe, which, by virtue of having nothing in it to move, was also, like Einstein's model, static.[28] De Sitter had further found that if a particle were introduced into his empty universe it would behave as if it possessed inertia, in violent and obvious contradiction to Mach's Principle. De Sitter termed his solution *Solution B,* to distinguish it from Einstein's, which he labelled *Solution A,* and the flat space–time of special relativity which constituted *Solution C* and which took no account of gravitational fields. The main characteristics of Solutions A and B were:

(1) Solution A can contain matter and remain stable, Solution B cannot.

(2) In the model corresponding to Solution A there is no systematic redshift, but Solution B indicated that the wavelength of light should increase – that is, shift towards the red – with increasing distance from the origin of the co-ordinates. The effect that de Sitter predicted was not due to a real recession of distant stars or nebulae. Instead the intrinsic properties of space and time in Solution B cause clocks to appear to run more slowly the further they are from the observer, and so the atomic vibrations within a far-off galaxy appear to slow down, the frequency of light decreases, the wavelength of light thereby increases and a redshift is observed.[29]

Einstein was shocked and puzzled by de Sitter's solution. In June 1917 he complained to de Sitter that the 'metric does not make sense to me. It could only apply without "world material", that is stars'.[30] De Sitter tried to escape from this apparently insuperable difficulty by arguing that his solution might be applicable to a universe in which the density is so low that a zero density is a good approximation to the actual situation.

While de Sitter did not advocate either Solution A or B in his 1916 and 1917 *Monthly Notices of the Royal Astronomical Society* papers, he did think that the presence of a K term in the radial velocities of the distant stars of two spectral types indicated that they were subject to a systematic spectral displacement of some kind, and hence provided a measure of observational backing to Solution B. But he suggested that a better test of his model would be an attempt to find a redshift–distance relation among the objects beyond the Galaxy since the more remote the object, the more clearly the redshift would reveal itself. As de Sitter supposed the spirals to be island universes, he expected them to exhibit such a relation if his model approximated the properties of the Universe.[31]

By 1917 Slipher possessed the radial velocities of 25 spirals, yet because of the difficulties in communication caused by the First World War, when de Sitter wrote his third paper he knew of only three of them, one of which was the large negative velocity of 300 km s^{-1} for the Andromeda Nebula. In de Sitter's opinion this anomaly could be explained away as a large peculiar motion superimposed on the small redshift to be anticipated from the redshift–distance effect, but even four years later, when he knew of 25 spiral radial velocities and the preponderance of recessional velocities had become plain, he was still wary of accepting Solution B fully.[32] His hesitancy was probably due to his doubts about the distances to the spirals: without knowing them accurately, de Sitter could hardly claim that the more distant had larger redshifts since he was not sure which spirals were in fact the nearer and which the further.

At first observational astronomers did not pin much faith on the models of de Sitter and Einstein. An acceptance of the general theory of relativity was a necessary condition for an astronomer to do so, and in the late 1910s and 1920s many of them were diffident, and sometimes hostile, towards the theory. One reason why the theory was viewed sceptically was its mathematical abstruseness, even after the results of the British eclipse expeditions in 1919 to Sobral and Principe had apparently vindicated Einstein's theory so sensationally by measuring the deflection of light as it passed the Sun.[33] Hale was probably speaking for many astronomers in 1920 when he admitted that 'the complications of the theory of relativity are altogether too much for my comprehension. If I were a good mathematician I might have some hope of forming a feeble conception of the principle, but as it is I fear it will always remain beyond my grasp'.[34] But even in the early 1920s Einstein's theory was being treated seriously and was a topic of debate, although many astronomers still viewed it with suspicion.[35] For example, in 1922 Lick Observatory organised an eclipse expedition to test the general theory by measuring the deflection of star-light as it passed by the Sun. When the measurements were complete, Campbell lamented that the agreement between the theory's predictions and the observations was as close as the most ardent proponent of general relativity could hope for. He had 'hoped it would not be true'.[36]

As general relativity became more widely discussed, so de Sitter's model of the Universe gained in plausibility and it stimulated a number of investigations. Much of this interest was a consequence of the writings of Eddington, especially his celebrated *The mathematical theory of relativity* of 1923. Here he reported:

It is sometimes urged against de Sitter's world that it becomes non-statical as soon as any matter is inserted in it. But this property is perhaps rather in favour of de Sitter's theory than against it.

One of the most perplexing problems of cosmogony is the great speed of the spiral nebulae. Their radial velocties average 600 km. per sec. and there is a great preponderance of velocities of recession from the solar system. It is usually supposed that these are the most remote objects known (though this view is opposed by some authorities), so that here if anywhere we might look for effects due to a general curvature of the world. De Sitter's theory gives a double explanation of this motion of recession; first, there is the general tendency to scatter ... second, there is the general displacement of spectral lines to the red in distant objects due to the slowing down of atomic vibrations which ... would be erroneously interpreted as a motion of recession.[37]

Eddington's first impression in 1917 had been to feel a strong objection to Solution A.[38] He dismissed the world matter of Einstein's universe as no better than the aether (that all-pervasive, intangible, invisible substance that classicial physics needed as the carrier of light waves). Furthermore, it was not even as useful as the aether since the world matter's function, Eddington objected, 'seems to be just to be there'.[39] In contrast, he thought Solution B 'promising'. But a year later, in 1918, he revealed to a correspondent:

De Sitter's hypothesis does not attract me very much, but he predicted this (spurious) systematic recession before it was discovered definitely; and if, as I gather, the more distant spirals show a greater recession that is a further point in its favour.[40]

Despite the fact that by 1923 he had done much to publicise de Sitter's solution, Eddington now preferred Einstein's. Although he realised that it was incapable of explaining the redshifts, Eddington perceived in Einstein's solution the chance of linking the ratio of the electron's radius and mass to the number of particles in the Universe, and for him this outweighed the advantages of de Sitter's solution.[41]

In summary, by the early 1920s astronomers recognised that general relativistic cosmology offered two models of the Universe: Einstein's Solution A and de Sitter's Solution B. Which of these admittedly crude models better explained the large scale properties of the Universe, or whether some kind of combination of them was required, were still open questions.

Early investigations of a redshift–distance relation

In his 1917 paper on the solutions to the field equations, de Sitter had proposed that if Solution B represented to a good approximation the actual properties of the Universe, then a relation between redshift and distance was to be expected: the greater the distance of a body, the larger its redshift. Indeed, the existence of such a relation was soon to be widely assumed, although its form was not agreed upon and it was contentious whether or not it represented the effect predicted by de Sitter.[42]

However, a formidable obstacle had to be overcome before securing a relation between redshift and distance that would be convincing to the majority of astronomers: the accurate determination of the distances to the spirals. But in 1923 and 1924 this barrier was in part removed when Hubble discovered Cepheid variables in nearby spiral nebulae. Before Hubble's observations of Cepheids, astronomers had been restricted almost entirely to the crude distance indicators provided by the novae, the apparent luminosities and the apparent diameters of extragalactic nebulae.[43] As Hubble recalled in 1936, a glance at a plate of the Andromeda Nebula and its two companion nebulae was enough to show how unsatisfactory the latter two indicators were. These three nebulae, because of their proximity in the sky, were assumed to be physically related and so at a similar distance; but their brightnesses and diameters differed greatly.[44] In addition, the absolute magnitudes of the novae that had been detected in a few spirals were hotly disputed, and so their distances uncertain.

Despite these objections, a few tries were made nevertheless to construct a redshift–distance relation using the blunt tools of novae, constant brightnesses and diameters. It is to these that we now turn, and the better to understand these efforts we shall first analyse the attempts of one Polish–American mathematical physicist to establish a relation for the globular clusters. Ludvik Silberstein employed the globular clusters as his test objects because he judged that as a result of Shapley's extensive researches, their distances were known more accurately than those of the spirals. The spectral shifts of the globular clusters, however, contained shifts to the blue and red in roughly equal numbers. This was not a problem for Silberstein because he had derived an expression for the relation between the wavelength shift and distance in which the fractional change in wavelength, $d\lambda/\lambda$, could be either (r/R) or $- (r/R)$ when (r/R) is small, where λ is the wavelength of radiation emitted, r the distance and R the radius of curvature of the universe. In 1924, he announced that he had found a definite correlation between spectral shift and distance.[45]

The possibility that the globular clusters exhibited a redshift–distance relation had already been mooted in 1920 by Shapley and Russell.

Russell had suggested to Shapley that the radial velocities of the spiral nebulae might be due to a curvature of the space–time manifold, and he cited as his authority de Sitter's 1917 paper.[46] Shapley, who, like Russell, at this time rejected the existence of visible island universes, replied that there was as yet no observational evidence for the de Sitter effect.[47] Since the globular clusters are so remote, at least as distant as the spirals, why do they not display such a relation? This convinced Russell: 'Your point about the relativity effect on the velocities of spiral nebulae and globular clusters is very illuminating. I think you have disposed of de Sitter's suggestion for the present.'[48]

In 1923, 1924 and 1925, Shapley also corresponded with Silberstein, and indeed encouraged him. Silberstein confided to Shapley early in 1924 that the studies of Eddington and Weyl on de Sitter's model were based on 'guesswork',[49] and he later sent Shapley a velocity–distance diagram on which were plotted the positions of the Magellanic Clouds and seven globular clusters (this was also printed in *Nature*).[50] Shapley was not as impressed with Silberstein's analysis as Silberstein might have been led to believe from Shapley's letters. Later in 1924, in a note to Eddington, Shapley remarked that he had been compelled to discourage one of his students' enthusiasm for the 'Silberstein effect'.[51]

Silberstein's papers on the redshifts were written in a polemical style and Shapley was not alone in distrusting their conclusions. The value Silberstein had calculated for the radius of curvature of the Universe was only about twice Shapley's estimated size for the Galaxy, and this flagrantly conflicted with Lundmark's belief in island universes. Lundmark responded by first citing the authority of Eddington and Weyl, both of whom believed that Solution B was capable of generating only velocities of recession.[52] Secondly, Lundmark had checked several classes of stars, such as Cepheids and O-stars, for signs of a noticeable spectral shift. If Silberstein's value for the radius of curvature of the Universe was correct, they should have disclosed such evidence; Lundmark decided they did not. But the most destructive shot Lundmark fired was aimed at Silberstein's selection of data. Why, Lundmark asked, had Silberstein inserted into his calculations only 7 out of 16 known globular cluster radial velocities? His own answer was that Silberstein had used only those globular clusters that gave a constant radius of curvature.

Silberstein's assertions also prompted Lundmark to investigate the possible existence of a redshift–distance relation amongst the 44 spirals for which he knew the radial velocities. Using the novae that had been detected in the Andromeda Nebula, Lundmark estimated its distance to be 200 000 pc and this he took as the basic unit of the distance scale, based on constant diameters and constant absolute magnitudes, that he manufactured to calculate the distances to the smaller spirals. Employing these distance indicators was, he admitted, 'a rather rough hypothesis, but probably the best we can use for the present'. With the aid of this 'rough hypothesis' he concluded tentatively that there may be a re-

Fig 20. Lundmark's 1924 velocity–distance diagram for the globular clusters and the Magellanic Clouds. From Lundmark (1924*b*) 753.

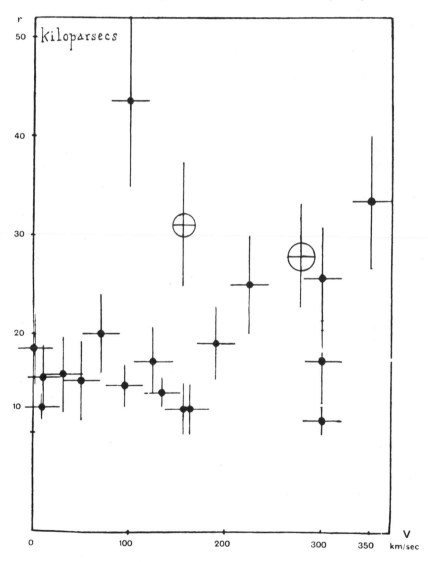

lation between redshift and distance, although 'not a very definite one'.[53]

The Mount Wilson astronomer Gustaf Strömberg, independently of Lundmark, also publicly criticised Silberstein for drawing inferences that were not justified by his data. While Strömberg's paper was written primarily to advance his own hypothesis of a fundamental reference frame in which there is an upper limit, well below the velocity of light, to the observed velocities of the celestial bodies, he had in addition analysed the radial velocities of 43 spirals, the Magellanic Clouds and 18 globular clusters. He found there was 'no sufficient reason to believe that there exists any dependence of radial motion upon distance from the Sun'.[54]

In their searches for a relation between radial velocity and distance both Strömberg and Lundmark had employed the standard method of calculating the solar motion and so had solved equations of the form

$$V = X \cos \alpha \cos \delta + Y \sin \alpha \cos \delta + Z \sin \delta + K.$$

In 1925, Lundmark made yet another determination of the solar motion and now he replaced the constant K term by a power series dependent upon r (the relative distance of the spirals derived from the assumption

Fig 21. Lundmark's 1924 velocity–distance diagram for the spirals. From Lundmark (1924*b*) 768.

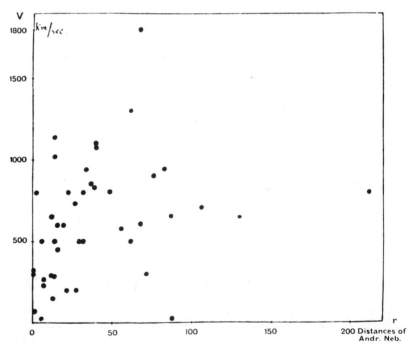

of constant diameters and constant absolute magnitudes for the spirals, with the Andromeda Nebula taken as the fundamental type). The solar motion equations now took the form

$$V = X \cos \alpha \cos \delta + Y \sin \alpha \cos \delta + Z \sin \delta + k + lr + mr^2,$$

where k, l and m are constants. Lundmark's derived expression for the power series was

$$K = 513 + 10.365\, r - 0.047 r^2\, \mathrm{km\ s^{-1}}.$$

The negative term, he admitted, was small and uncertain, but he was convinced that it was genuine and not the spurious product of meagre data. Its presence implied an upper limit to the velocity of the spirals, and since the maximum value of the K term would be roughly $2200\,\mathrm{km\,s^{-1}}$, to which could be added a maximum peculiar velocity of about $800\,\mathrm{km\,s^{-1}}$, this upper limit would be around $3000\,\mathrm{km\,s^{-1}}$. Although Strömberg had suggested that the K term is probably much larger for NGC 584 and smaller for the Andromeda Nebula than for the rest of the spirals,[55] he had not seen the possibilities of his suggestion. Hence it was left to Lundmark to make the explicit introduction of a K term dependent upon distance.[56]

Lundmark's innovation was not picked up at once. In 1926, Dose, a colleague of Wirtz at Kiel, wrote a paper in which he still followed the orthodox method of calculating the solar motion from the radial velocities of the spirals. While Dose remarked that the redshifts of non-galactic objects seemed to increase as the diameters decreased, he did not attempt to link this explicitly to a variable K term.[57]

During the 1920s a number of 'anticipations' of a linear redshift–distance relation were made by investigators of the mathematical properties of the de Sitter model. In 1923, Hermann Weyl, a brilliant German physicist and mathematician and author of the highly influential *Welt, Raum, Zeit*, had combined de Sitter's solution and the hypothesis that the stars lie on a pencil of geodesics that diverge from a common event in the past (this became known as Weyl's Principle[58]), and thereby arrived at the expression $(d\lambda/\lambda) = \tan (r/R)$ for the displacement of the spectral lines of a distant body.[59] For small (r/R) this reduces to $(d\lambda/\lambda) \simeq (r/R)$, where r is the measured distance of the star in the static space at the moment that the observation takes place. Hence, to a first approximation, Weyl's analysis indicated a linear relation between redshift and distance. Another such prediction was published in the *Annales de la Société Scientifique de Bruxelles* of 1927.[60] The author was Abbe

Georges Lemaître. Lemaître had sought, and he believed had discovered, a solution to the field equations that combined the virtues of Solution A and Solution B; that is, a solution that allowed the Universe to contain matter, but that could account for the redshifts of the galaxies. The radius of curvature of Lemaître's model increased with time, and for the redshifts of (cosmologically) nearby objects (redshifts which he took to be genuine radial velocities) he derived $(d\lambda/\lambda) = (r/R_0\sqrt{3})$, where R_0 is the radius of curvature of the Universe at the time of emission. The following year, in 1928, the mathematician H. P. Robertson derived an approximate relation between the velocity of a galaxy and its distance: $V \sim (cr/R)$. In this expression r is the galaxy's distance and c is the velocity of light. Robertson, too, had coupled de Sitter's solution and Weyl's principle, and by transforming co-ordinates he obtained what he and his contemporaries viewed as a cosmology mathematically equivalent to de Sitter's. In addition, Robertson compared the distances of the galaxies given by Hubble in 1926 with Slipher's list of radial velocities, and concluded that he had achieved a 'rough verification' of $V \simeq (cr/R)$ and a value of $R = 2 \times 10^{22}$ km;[61] hence, the redshift implied by the de Sitter model was 'in accord with known facts concerning the radial velocities of spiral nebulae'.[62] The analyses of Weyl, Lemaître and Robertson, however, excited little interest among astronomers.

Hubble's 1929 paper

In 1926, at the end of his paper on 'Extra-galactic nebulae', Hubble had expressed publicly his interest in de Sitter's Solution B, although the section he devoted to 'The finite universe of general relativity' is a little confused and reveals that he had not appreciated all of the differences between the Einstein and de Sitter models.[63] By 1928 Hubble had a clearer idea of the properties of Solution B, especially the prediction of a redshift–distance relation. Also in 1928, the first phase of Hubble's exploration of extragalactic space was drawing to a close with his lengthy paper on the Andromeda Nebula (discussed in Chapter 3), and he was ready to begin a new line of investigation. After attending the 1928 International Astronomical Union Meeting in Holland, a meeting at which Hubble had had a chance to discuss nebular problems with many of the leading practitioners in the field, Hubble chose to tackle the puzzle of the redshifts, and in particular he decided to test de Sitter's solution.[64] Certainly such a choice was a completely natural development from his earlier studies since the predominantly recessional radial velocities of the nebulae constituted one of the oustanding anomalies in extragalactic astronomy.

By the mid-1920s Slipher's gallant assault on the radial velocities of the spirals had come to an end. Slipher had made his measurements with a fine short focus spectrograph attached to a 24-inch refractor. For the brighter and bigger nebulae this was an admirable combination since for the large, fairly uniform surfaces which these objects presented, all telescopes were almost equally effective because the brightness of the spectrum was not increased by a change in the aperture or focal length.[65] The spectrograph was the crucial instrument because it fixed the speed of observation. However, with the smaller and fainter nebulae the light grasp of the telescope became increasingly important, and by the early 1920s Slipher was probing so far into space and looking at such small nebulae that the Lowell 24-inch refractor was proving more and more of a handicap. So when in 1928 Hubble wanted to measure the redshifts of extragalactic nebulae, he turned to Milton Humason and the 100-inch telescope at Mount Wilson. After 1928 Mount Wilson became the centre for such measurements and Humason, a meticulous and highly gifted observer, assumed Slipher's role.[66]

When in 1929 Hubble described his first researches on the redshifts, he possessed values for 46 extragalactic nebulae and he claimed accurate distances to 24 of them.[67] When he plotted the redshifts of these 24 nebulae against their distances, he judged that a linear redshift–distance relation was the simplest way of representing his data. In order to incorporate this finding, he changed the solar motion equations to

$$V = X \cos \alpha \cos \delta + Y \sin \alpha \cos \delta + Z \sin \delta + kr.$$

This was the standard equation for calculating the solar motion except for one notable innovation: he had replaced the constant K term with a term linearly dependent upon distance. Provided the redshifts were really Doppler shifts, this equation implied a systematic recession of the extragalactic nebulae represented by $V = kr$. Hubble analysed his data in two ways. First, he calculated the solar motion and the K term from the radial velocities of the 24 nebulae whose distances he was confident that he knew accurately. Secondly, he put these 24 nebulae into 9 groups according to proximity and direction and reworked the calculations. The two sets of solutions agreed very closely and for 'such scanty material, so poorly distributed, the results are fairly definite'[68] that a linear relation does exist between the redshift and distance. He next examined the 22 nebulae for which he knew the distances with less certainty: the results again supported the existence of a linear velocity–distance relation. Hubble also drew a velocity–distance diagram in his paper; its superiority over previous diagrams is obvious since the fit is tighter and the distances were, at least in Hubble's view, determined more precisely.

Hubble had planned his research in order to test de Sitter's model of the Universe, and he declared that the major result of his investigation was the 'possibility that the velocity–distance relation may represent the de Sitter effect, and hence that numerical data may be introduced into discussions of the general curvature of space'.[69] (We should stress that in the term 'de Sitter effect' Hubble in fact included the two effects in de Sitter's model that together caused the phenomenon of an increasing redshift with increasing distance: (1) the apparent slowing down of the vibrations of distant atoms and (2) the scattering of particles in the de Sitter world.[70]) He also believed that he was conducting a critical test that would soon allow him to dismiss either Solution A or Solution B: 'The necessary investigations are now under way with the odds, for the moment, favoring de Sitter.'[71] Yet in his 1929 paper Hubble made no reference to any theorist who had influenced his research programme apart from de Sitter. He may have done this in order to emphasise the observational foundations of his relationship since Hubble throughout his career was somewhat suspicious of theory. Also, Hubble may have shied away from discussing extensively the theoretical basis of his study because he

Fig 22. Hubble's velocity–distance plot of 1929. From Hubble (1929*b*, 172). The black discs and full line represent the solution for solar motion using the nebulae individually, the circles and broken line represent the solution combining the nebulae into groups.

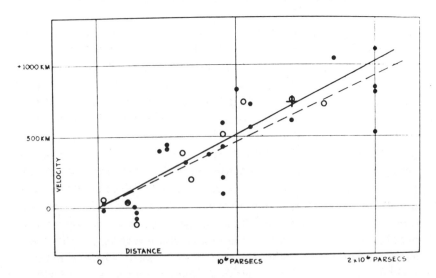

felt (as N. S. Hetherington has cogently argued) that a stigma was still attached to the pursuit of a redshift–distance relation as a result of Silberstein's extravagant claims of the mid-1920s. It was probably in an effort to ensure that this stain would not prejudice the reception of his own investigation that Hubble asked Strömberg to check his solar motion solutions by selecting groupings of the nebulae different from his own. 'Certainly Hubble did not need Strömberg', Hetherington has noted, 'to check the limited and relatively straightforward calculation of the solar motion'.[72] Rather, because Strömberg had rebuked Silberstein, Strömberg's name was added to Hubble's paper to allay the doubts of those astronomers who had been disturbed by Silberstein's papers. Hubble also quoted Lundmark's value for the solar motion. When we remember that Hubble was generally wary of Lundmark's results (in 1926 Hubble had even accused Lundmark of plagiarising his scheme of nebular classification), we can only agree with Hetherington that Hubble mentioned Lundmark because he too had attacked Silberstein.

By 1929 a redshift–distance relation was not unexpected; indeed, in 1930 Hubble himself told de Sitter that 'the possibility of a velocity–distance relation among nebulae has been in the air for years – you, I believe, were the first to mention it'.[73] As we have seen, during the 1920s it was not clear what form the relation took and numerous attempts had been made, mostly stimulated by de Sitter's Solution B, to determine its nature observationally. The thread linking these efforts was the use of the solar motion equations and the K term which, after Paddock's use of it in 1916, had become a standard tool. Also, during the 1920s a number of mathematical researchers had investigated what form of relation was to be expected. Hence, Hubble was working within a well-defined problem area, using a well-defined method. Furthermore, his computational scheme involved only a minor change from the usual solar motion equations: he wrote kr instead of K. This move, as apparently innocent but as far-reaching as Copernicus's removal of the equant point, was the fundamental step which when allied to a set of bold mathematical researches led to a new conception of the Universe.

The reception of Hubble's relation

The reality of Hubble's linear redshift–distance relation was quickly accepted by many astronomers, and the only serious resistance came from Harlow Shapley. Shortly after reading Hubble's 1929 paper Shapley composed what he saw as a rebuttal of Hubble's claimed linear relation, but even Shapley admitted there was a correlation of some form between the velocity and distance of a nebula. Maybe he was in part regretting a lost opportunity to pursue such a relation himself because in his paper Shapley first recalled his remark of 1919 that the spiral nebulae recede from both sides of the Galaxy and that the rate of this recession might depend on velocity or mass. Now in 1929, he warned that although the additional data secured in the intervening years had provided 'some further support to the hypothesis of a dependence of measured velocity on apparent magnitude and distance . . . the extension to fainter objects depends on only a few spectrograms'.[74] In addition, Shapley pointed to what he saw as flaws in Hubble's carefully constructed distance scale. Shapley wrote that the large dispersion in the intrinsic brightness of the galaxies made their integrated absolute magnitude a 'treacherous indicator of relative distance; the same uncertainty attends the use of angular diameters, and also, to a lesser extent, the use of the brightest stars'.[75] Except for those six extragalactic nebulae in which he had detected Cepheids (and other types of stars, such as novae, were used as a check on the distance indicated by the Cepheids), Hubble's distances to what he took to be the nearer nebulae depended upon measuring the apparent luminosity of the brightest stars. Then, by comparing the apparent luminosity with the estimated intrinsic luminosity of these stars, the distances to the nebulae containing them could be calculated. But, Shapley argued, in a galaxy at a great distance a cluster will be indistinguishable from a stellar image. For example, if the Pleiades star cluster were placed in the Coma–Virgo cluster of galaxies it would have the appearance of a star relatively near to us.[76]

Before writing this paper Shapley had corresponded with Hubble, and in May 1929 Shapley queried a number of inconsistencies he had noticed between Hubble's quoted radial velocities and other published values (an issue he was to raise in his paper).[77] In particular, why was the Andromeda Nebula given a velocity now of $-220\,\mathrm{km\,s^{-1}}$, when all previous determinations were around $-300\,\mathrm{km\,s^{-1}}$? Shapley's implication was obvious: if one is so unsure about the radial velocity of such a nearby galaxy, what confidence could one have in those of more distant, and fainter, galaxies? Hubble replied that the value of $-220\,\mathrm{km\,s^{-1}}$ for the Andromeda Nebula had been secured from a very fine plate centred on the nucleus, but he suspected that the velocity was larger outside the nucleus and perhaps this explained why earlier investigators had found a velocity of $-300\,\mathrm{km\,s^{-1}}$. The usually cautious Hubble ventured the bold suggestion that this difference was as if 'there were a relativity shift to the red in the nucleus itself'. As to the existence of a linear relation, Hubble told Shapley:

My paper, will you realize, is merely a preliminary correlation of the data available and makes no claims to finality. In a few years we should have sufficient new data to re-examine the question in a comprehensive manner. I believe that a relation will still be found but whether it will be linear is perhaps an open question.[78]

Even now Hubble confided that he would have preferred to delay publication, but he was concerned that as soon as astronomers heard of the high velocities Humason was finding some of them would jump unfairly into what he believed to be his field; as 'it is, the paper was held for over a year'.[79]

Shapley described the contents of Hubble's letter to Russell. He believed that Hubble was less sure of the linear relation and that Slipher's assault on the spirals had 'stopped just short of some great excitement'.[80] However, even Shapley's opposition to the linear relation soon waned and in April 1930 he reported that 'Hubble has presented evidence indicating, in the first approximation, that there is a linear relation between distance and speed – the more remote a galaxy, the faster its recession.'[81]

There were four main reasons for the rapid acceptance of the linear redshift–distance relation and we shall discuss each of these in turn. First, a number of astronomers had attempted before 1929 to establish observationally a relation between distance and redshift, and so the ground that Hubble had to cover was already well staked out. Hubble's study was, in the eyes of his colleagues, superior to these earlier attempts because he had used distance finding techniques that were usually accepted as providing the most accurate values available. As we have seen, Hubble had also taken great care to distance himself from Silberstein's studies. Secondly, Hubble had employed the world's most powerful telescope, the 100-inch reflector at Mount Wilson, and his authority in extragalactic astronomy also argued for the reliability of his studies of extragalactic nebulae and hence to the correctness of the redshift–distance relation to which they had led. Thirdly, the main conclusion of Hubble's 1929 paper was swiftly corroborated by de Sitter who in 1930 also derived a linear redshift–distance relation from the observations available to him. De Sitter noted that it 'has been remarked by several astronomers that there appears to be a linear correlation between the radial velocities and the distances' of external galaxies,[82] and he agreed. De Sitter's comments, coming as they did from one of the most eminent and influential astronomers, added credence to the reality of the linear relation.[83] Fourthly, the relevance of the linear relation for models of the Universe was soon recognised. By the late 1920s mathematical cosmologists had become acutely aware of the inadequacies of the static solutions, Solution A and Solution B. In 1929 R. C. Tolman voiced this concern when, after rejecting Einstein's solution, he contended that the de Sitter solution *'does not afford a simple and unmistakably evident explanation of our present knowledge of the distribution, distances, and Doppler effects for the extragalactic nebulae'*.[84] When de Sitter addressed the Royal Astronomical Society in January 1930 he too expressed his doubts about Solutions A and B. As part of his talk he showed a slide of a linear relation that he had derived using a distance scale based on the assumptions of the uniformity of diameters and brightnesses of the galaxies. By combining the velocity–distance relation and Solution B he had calculated that $R = 2 \times 10^9$ light years, which, to his dismay, was roughly the same as the value he obtained if he instead used Solution A as the basis of his calculation. 'I am unable', de Sitter confessed, 'to account for this coincidence'.[85] The coincidence implied to de Sitter that the mean density of matter was too high for Solution B to be a good approximation to the observed universe, since Solution B could only be applicable to a universe in which the mean density of matter was close to zero. But he also felt that the alternative, solution A, was equally unsatisfactory because it was completely unable to account for the redshifts of the galaxies. Moreover, in 1922 the Russian meteorologist and mathematician Alexander Friedmann had proved[86] (and the proof of the same result had been brought before a wider audience in 1929 by Tolman[87]) that there are only three solutions to the field equations for the metric of a static universe: Solution A, Solution B and the solution of special relativity (which took no account of gravitational fields). By early 1930 it seemed that the three were all but worthless as representations of the actual Universe.

In this atmosphere of crisis Lemaître's paper of 1927 was 'discovered'. If Lemaître's paper had been neglected earlier, in 1930 it was welcomed enthusiastically, and was soon widely seen as underpinning the hypothesis that the redshifts are the product of an expansion of the Universe.

The expanding universe
At the end of de Sitter's talk to the Royal Astronomical Society in January 1930, Eddington had remarked that he was puzzled as to 'why

there should be only two solutions [A and B]. I suppose the trouble is that people look for static solutions.'[88] Eddington had forgotten about Lemaître's paper. However, upon reading Eddington's comments Lemaître, an ex-student of Eddington's, wrote to remind Eddington of his paper of three years earlier. Eddington, who at the time was working on the stability of the Einstein model by analysing Robertson's solution, now recognised immediately the value of Lemaître's study and he was delighted to find that it allowed him to escape from the dilemma of having to choose *either* Solution A *or* Solution B by choosing neither.[89]

De Sitter also greeted Lemaître's study warmly. Just before Eddington had told him of Lemaître's paper, de Sitter had felt compelled to conclude that '*both the solutions* (A) *and* (B) *must be rejected,* and as these are the only statical solutions of the equations . . . the true solution represented in nature must be a dynamical solution'.[90] It seems that by 'dynamical' he really meant 'non-static' because in April 1930 de Sitter revealed to Shapley:

I have been very busy lately on spiral nebulae and on the relativistic explanation of the big velocities. I had come to the conclusion that my Solution 'B' could not be accepted as an adequate explanation, as it supposes the universe to be empty, or at least emptiness to a good approximation, and the actual density of spirals is so large as to make it nearly full, (as it should be in Einstein's solution 'A', which, however does *not* admit large systematic velocities). Only very lately I have found the true solution, or at least a possible solution, which must be somewhere near the truth, in a paper . . . by Lemaître . . . which had escaped my notice at the time.[91]

In 1931 de Sitter went further and announced that the solution to the puzzle of accepting either Solution A or B had been provided by Lemaître, whose 'brilliant discovery, the "expanding universe", was discovered by the scientific world about a year and a half ago, three years after it had been published'.[92] (De Sitter's claim that Lemaître discovered the 'exanding universe' is incorrect, as we shall see.)

Einstein had taken little interest in cosmological matters after his 'Kosmologische Betrachtungen' paper of 1917, but in early 1931 he visited Pasadena and called at the Mount Wilson offices. The day after he arrived Einstein made known his rejection of a static universe:

New observations by Hubble and Humason . . . concerning the redshift of light in distant nebulae make the presumptions near that the general structure of the Universe is not static.
Theoretical investigations made by Lemaître and Tolman . . . show a view that fits well into the general theory of relativity.[93]

Einstein no longer had need of the Λ term, which he had originally introduced to produce a static solution, and he quickly removed it from his
field equations.[94]

While non-static solutions to the field equations made no 'sudden and
dramatic appearance' in cosmology,[95] it was not until the late 1920s that
there seemed to be any pressing reason for introducing such a solution.
In consequence, solutions such as Friedmann's of 1922[96] in which the
space curvature depended only on time, had provoked little interest
when they were first expounded.[97] It was not realized during the 1920s
that it was a mathematical accident that the de Sitter metric was
expressed in a time-independent form. As G. J. Whitrow has written:
'What de Sitter had in fact discovered in 1917 was one of the simplest
models of an expanding universe. With a more physically appropriate
choice of co-ordinates, as was first shown by G. Lemaître in 1925, and
independently by H. P. Robertson in 1928, the metric of the de Sitter
universe can be expressed [in the form of] the limiting case of an expanding universe as the mean density everywhere tends to zero. We therefore

Fig 23. H. N. Russell (left) with de Sitter (far right) and Shapley
(second from right) *c.* 1930 (Courtesy of A.I.P. Niels Bohr Library,
Margaret Russell Edmondson Collection).

no longer consider de Sitter's as a static universe, its apparent change-lessness being a mathematical fiction.'[98]

For many, a belief in the existence of the linear redshift–distance relation was linked to an acceptance of the redshifts of the galaxies as being caused by their radial velocities. Hence, to these astronomers and mathematicians the relation implied an expanding universe, but they soon realised that this did not indicate a uniform expansion of space. Rather, it was argued that it is the space *between* the galaxies that expands. Since the gravitational fields within a galaxy are many more times greater than those outside, the formulae applicable to the Universe can not be applied within a galaxy.[99]

In 1931 the existence of a linear redshift–distance relation was further strengthened by the promised sequel to Hubble's 1929 paper. The use of a linear relation had appeared to Hubble in 1929 to be the simplest way available of representing his data, but he may have been inclined towards such a form of the relation because of the then current theoretical predictions.[100] On the other hand, in the last paragraph of his paper Hubble wrote that the linear relation was a 'first approximation representing a restricted range in distance',[101] and so Hubble suspected that it might be the tangent at the origin to a more general relation.[102] De Sitter had originally proposed a relationship such that $(d\lambda/\lambda) \propto r^2$; but because it had become widely accepted in the 1920s that material particles scatter in the de Sitter world, there was great confusion about the exact form of the relation predicted by the de Sitter model.[103] Although Hubble was confident that he had established a linear relation out to 2×10^6 pc, he wanted to confirm that it applied for more distant objects. In 1936, in his classic *The realm of the nebulae,* Hubble explained the strategy behind his public methodology:

[An] isolated group of data is studied and the results are interpreted against the background of general knowledge. Then follows the process of extrapolation, and tests, and appropriate revision. The observations and the laws which express their relations are permanent contributions to the body of knowledge ... The research sweeps outward and develops an observable region around a given center – a realm of positive knowledge. Beyond the horizon, is the realm of speculation. The observer, if he ventures therein, can only throw his empirical relation into the blue, and search for inconsistencies with extrapolations from other centers.[104]

Thus in 1929 he was resolved to throw his linear relation 'into the blue', with himself securing the distances to the extragalactic nebulae and Humason the radial velocities. The outcome was Hubble and Humason's paper of 1931 entitled 'The velocity–distance relation among extra-galactic nebulae'.[105] The two authors started their paper with a vigorous defence of the procedures they had followed to obtain the distances of remote extragalactic nebulae, probably in response to Shapley's criticisms of 1929. The first step in these measurements was the identification in the nebulae of particular types of stars whose absolute magnitudes were regarded as known. By then comparing the apparent magnitudes with the known absolute magnitudes, the distance to a nebula containing these stars could be calculated. By 1931, Hubble and Humason had identified in eight nebulae at least one of the following types of star: Cepheids, P Cygni stars, Novae, irregular variables, and helium stars. Using these as distance indicators Hubble and Humason advanced out to a distance of 7.3×10^6 pc. Also, in over half of those nebulae whose radial velocities had been determined, they observed star-like images that they claimed were the most luminous stars in the nebula. After calibrating these stars with the most luminous stars in systems whose distances had been measured, Hubble and Humason could derive the distances to another 30 nebulae. The last link in this chain was secured by

Fig 24. Hubble and Humason's 1931 velocity–distance diagram. From Hubble (1931) 77.

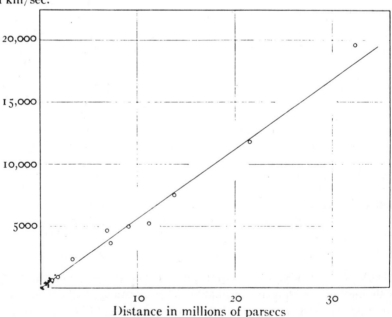

Velocity in km/sec.

Distance in millions of parsecs

the apparent magnitudes of the nebulae (which Hubble now believed were more reliable as distance indicators than apparent diameters): the absolute magnitudes of the relatively nearby nebulae were known, and so, by comparing with the apparent magnitudes, the distance to more distant nebulae could be calculated.

By exploiting these distance criteria Hubble and Humason drew a revised diagram of redshift versus distance. Now the remotest nebula on the diagram lay at an estimated 32×10^6 pc, an enormous 'leap into the blue' when contrasted with the most distant object at 2×10^6 pc in the 1929 plot. There were, in addition, many more points on the 1931 diagram because 40 more radial velocities were available. This rapid extension of the relation was made possible by the combination of an 'extraordinarily efficient'[106] and very fast spectrograph objective of ratio f/0.6 on the 100-inch telescope, and Humason's skill as an observer.

Despite the value of K being changed in the 1931 paper from 500 to $558 \, \text{km} \, \text{s}^{-1} \, \text{Mpc}^{-1}$, primarily because Shapley had recalibrated the period–luminosity relation of the Cepheids,[107] the linear correlation between redshift and distance appeared to be even clearer than before.[108]

Although, as we have noted, many astronomers and mathematicians accepted that the redshift–distance relation signified an expansion of the Universe, others were incredulous at the size of the 'velocities' Humason was finding: if the redshift of one galaxy was interpreted as Doppler shift, then it was receding at an astonishing $19\,700 \, \text{km} \, \text{s}^{-1}$.[109] The value of the constant in the relation also seemed – if one simply extrapolated the present expansion back in time and took no account of a possible acceleration in the expansion – to imply that there was a time about 2×10^9 years ago when all the material in the Universe had been packed together much more closely than now. From here it was a short step to talking about of the 'age' of the universe. For those who accepted Jeans's estimate of 10^{13} years as the age of the Galaxy[110] (a value in accord with contemporary estimates of the time scale of stellar evolution), a time of 2×10^9 years was far too recent. How, they asked, could the age of the Galaxy be so much greater than the age of the Universe? Nor was the adopted value of the Hubble constant free from criticism. In 1932 Eddington remarked that many astronomers 'would adopt a value higher for the Hubble constant than 550 $[\text{km} \, \text{s}^{-1} \, \text{Mpc}^{-1}]$, believing that Hubble's scale of distances of the nebulae is systematically too great'.[111] Eddington, one of the strongest and most persuasive advocates of the expanding universe, contended that apart from those distances determined by the Cepheids, the distances assigned to the extragalactic nebulae were somewhat risky estimates, even though he felt they were not entirely misleading. Moreover, the higher the value of the Hubble constant, then the shorter the 'age' of the Universe and the more acute the time scale problem became. Nevertheless, criticism of the value of the Hubble constant was not often forthcoming: 'it appears at times that astronomers were more concerned with bringing the rest of astrophysics into line with the parameter in Hubble's Law than the other way about'.[112] Possibly the enormous authority conferred by the 100-inch reflector on Hubble and Humason's studies and Hubble's hard earned reputation as the pre-eminent extragalactic astronomer inhibited criticism.

In his 1929 paper Hubble had described as the outstanding feature of his study the possibility that the velocity–distance relation might represent the de Sitter effect, and he had ascribed the phenomenon to two causes: (1) a general scattering and (2) the properties of space–time in de Sitter's model that produce an apparent slowing down of the rate of distant clocks. At this time Hubble did not accept that a nebula's redshift was merely the product of its radial velocity, and we will recall that in 1929 Hubble had told Shapley that the measured redshift of the Andromeda Nebula might be due in part to a relativity effect.[113]

Between 1929 and 1931 astronomers debated the time scale problem, and when in 1931 Humason announced the results of his and Pease's measures of radial velocities, he gave his paper the title 'Apparent velocity-shifts in the spectra of faint nebulae'. Hubble likewise strove to avoid committing himself strongly to any one theoretical interpretation, as he made evident in 1931 to de Sitter:

[Humason and I] use the term 'apparent' velocities in order to emphasise the empirical features of the correlation. The interpretation, we feel, should be left to you and the very few others who are competent to discuss the matter with authority.[114]

The sort of scepticism about the nature of the redshifts that Hubble expressed was felt by many astronomers and spawned several attempts to find an alternative to the redshifts as pure Doppler shifts. None of them, however, gained much support. For example, in 1929 Fritz Zwicky of the California Institute of Technology argued that a gravitational analogue of the Compton effect, in which the photons of light on their journey to the Earth from distant galaxies transfer momentum and energy to gravitating material, would cause a loss of energy, and thereby produce a redshift.[115] And 'with a few plausible assumptions concerning the distribution of gravitating matter in intergalactic space a law not unlike Hubble's was, to Zwicky's satisfaction, explained'.[116]

De Sitter also disliked the 'naïve' interpretation of the velocity–distance relation. The beginning of the Universe and the start of its expansion, he suggested, could not be identified as the same event. The galaxies had existed for longer than 10^9 or 10^{10} years, but they had approached to a minimum separation before starting their expansion.[117]

Nevertheless, the critical, and often uncritical, acceptance of the bare mathematical concepts involved allowed astronomers and mathematicians to talk about the 'age' of the Universe. This led inevitably to the question of when the material content of the Universe was created.[118] By the early 1930s such a question had become an urgent one for a few cosmologists. No longer was this a question of metaphysics, a region to which nineteenth century science would have banished it. Even in 1929, when Hubble's paper was just off the press, Jeans had emphasised,

[contemporary astronomy] makes it clear that the present matter of the universe cannot have existed for ever; indeed we can probably assign an upper limit to its age of, say, some such round number as 200 million million years. And, wherever we fix it, our next step back in time leads us to contemplate a definite event, or series of events, or continuous process, of creation of matter at some time not infinitely remote.[119]

Conclusion

Although in the late 1910s and 1920s numerous investigations of a redshift–distance relation had been undertaken, it was not until 1929 that astronomers were assured by Hubble that such a relation really existed and that, at least to the first approximation, it was linear. The establishment of the relation had been made possible by the triumph of the island universe theory in the mid-1920s, and during the years between 1923 and 1931 the view astronomers held of the Universe had indeed expanded: in 1923 it had still been possible to argue convincingly that there are no visible external galaxies; by 1931 this was certainly not so. Moreover, by 1931 these galaxies were widely believed not merely to be visible, but to be actually dispersing.

The agreement on the reality of the redshift–distance relation also ushered in a new era in cosmology by presenting a novel set of problems, problems that had been largely dictated by the invention of the expanding universe. Further, the very character of cosmological investigation had been transformed. Hubble's relation and Lemaître's 1927 paper produced a dialectic between theory and data that was undreamt of only a few years before, and, although pitifully meagre by present day standards, it was to breed a confidence among astronomers and mathematicians that they could discuss, and ultimately explain, the entire history of the Universe.

Notes

1 Struve & Zebergs (1962) 469.
2 Whitrow (1972) 531.
3 For a full explanation of this method see Smart (1962) 263–75.
4 Truman (1916).
5 R. Young & Harper (1916).
6 E. Hertzsprung to A. Eddington, 18 May 1916, Aarhus and A. Eddington to E. Hertzsprung, 13 June 1916, Aarhus.
7 V. Slipher to W. Campbell, 18 December 1916, Lick.
8 Slipher (1917).
9 Slipher (1915) 21.
10 Campbell (1913) 208. The K term for the spirals was to be often referred to in the literature as the 'Campbell K term'.
11 Paddock (1916). Some writers have mistakenly claimed that C. A. Wirtz was the first to introduce the K term into these studies: see Hubble (1936) 107 and North (1965) 142.
12 Wirtz (1918).
13 See also Fernie (1970) 1221. A clear indication that Wirtz's proper motions of small spirals were spurious was that they did not depend on the magnitudes or diameters of the spirals: that is, they did not correlate at all with the distances of the spirals (Lundmark (1920) 75).
14 Lundmark (1920) 24.
15 Wirtz (1921).
16 Einstein (1917).
17 McCrea (1979) 756.
18 Einstein's use of 'curved' space to explain the large scale properties of the Universe was not novel (North (1965) 72–81).
19 North (1965) 72. The Einstein universe has the sections $t =$ constant as 3-dimensional spheres, while the 4-dimensional world is cylindrical. Professor W. H. McCrea has suggested to me that Einstein did not fully realize the possibility of doing without boundary conditions. Einstein's model had no boundary in space, and he avoided boundary conditions in time by *assuming* a steady state.
20 Moszkowski (1921) 127.
21 In 1929 Jeans asked the rhetorical question: 'Are there any limits to the extent of space?' Even a 'generation ago, I think most scientists would have answered in the negative. They would have argued that space could be limited only by the presence of something which is not space' (Jeans (1929) 70). General relativity, he proclaimed, had removed this objection and had laid open the way to the acceptance of a finite universe.
22 Einstein (1917) 188.
23 Einstein (1919).
24 Mach (1902) 229–38.
25 J. T. Blackmore writes that Einstein had 'couched the essential insight

of his theory of general relativity, the equivalence of inertial and gravitational mass, within a number of Machian ideas. Indeed, he considered the theory as a natural consequence of Mach's earlier work and believed that except for accidental factors Mach quite likely would have discovered and elaborated the theory himself (Blackmore (1972) 254). In June 1913 Einstein told Mach that from the 'basic and fundamental assumption of the equivalence of the acceleration of the reference frame and of the gravitational field' it is a necessary consequence that 'inertia has its origin in a kind of mutual interaction of bodies' (quoted in Blackmore (1972) 254).

26 A. Eddington to W. de Sitter, 11 June 1916, Leiden, and A. Eddington to W. de Sitter, 4 July 1916, Leiden. In his June letter Eddington had commented: 'Hitherto I had only heard vague rumours of Einstein's new work [on the field equations]. I do not think anyone in England knows the details of this paper.' Eddington was a quaker and his religious convictions compelled him to be a pacifist during the First World War. This meant he was free to pursue his scientific studies.

27 Kahn & Kahn (1975).

28 de Sitter (1917).

29 Professor G. J. Whitrow pointed out to me that the de Sitter model was the first that contained a time-horizon (or event-horizon in current terminology). That is, there is some finite distance, say R, from which an observer at the origin could never have any information of events occuring at R or beyond. Hence one can speak of an horizon to the Universe at R. Now another observer with a different origin would place his horizon differently, and so the spatially closed nature of the model is not special to any single observer. See also Tolman (1934) 354.

30 Kahn & Kahn (1975) 453.

31 Einstein did not think that island universes had been sighted. Writing to de Sitter in 1917, he remarked that one could see to a distance of 10 000 light years (quoted in Kahn & Kahn (1975) 452). In 1920 he was to comment that 'it is possible, in fact, to a certain degree probable, that we shall discover new worlds far beyond the limits of the region so far investigated' (Moszkowski (1921) 130). Perhaps he meant by the 'region so far investigated' the Galaxy, and by 'new worlds' external galaxies.

32 de Sitter (1922).

33 Douglas (1956) 39–43. At the island of Principe in the Gulf of Guinea the weather was poor and so the Sobral results were of greater weight. Typical of the excitement caused by the findings was Shapley's remark to a correspondent: 'What do you think of the principle of relativity now that the extreme case of the Einstein theory has been proved by the English eclipse results?' (H. Shapley to H. Benioff, 7 December 1919, Harvard).

34 G. Hale to Dyson, 9 February 1920, quoted in Clark (1973) 238.

35 For example, in 1924 C. V. L. Charlier, a leading Swedish astronomer, told an audience at the University of California that there are 'speculative men in our time who put the question whether space itself is finite or not, whether space is Euclidean or curved (an elliptic or hyperbolic space). Such speculations lie within the domain of possibility, but are of the same nature as the philosophical speculations ... in Kant's [*Critique of pure reason*]. They must be discussed on the

support of facts and at present they must be considered as lying outside such a discussion' (Charlier (1925) 182).

36 Quoted in Douglas (1956) 44.
37 Eddington (1923) 161. An example of this increased interest was the letter Hertzsprung sent to J. H. Reynolds in May 1923. Hertzsprung was excited because the 'circumstance, that the only negative "radial velocities" found for spiral nebulae belong to the two largest . . . just invites [us] to examine still fainter objects of this kind. I am pretty convinced, that the spiral nebulae have really great velocities by themselves. The question is for me, if there is superposed a de Sitter-effect in the displacements observed of the spectral lines' (E. Hertzsprung to J. Reynolds, 2 May 1923, Aarhus). Also, in reviewing *The mathematical theory of relativity,* Jeans had remarked that 'We are still far from able to assert that we live in a De Sitter universe, but the probability seems to increase steadily' (Jeans (1923b) 193).
38 A. Eddington to W. de Sitter, 16 August 1917, Leiden.
39 *Ibid.*
40 A. Eddington to H. Shapley, 30 December 1918, Harvard.
41 Eddington (1923) 168.
42 See Shapley & Shapley (1919) 126, Wirtz (1921) 349, Wirtz (1924) and van Maanen (1922b) 215.
43 On the use of novae as distance indicators see chapter 1.
44 Hubble (1936) 110.
45 Silberstein (1924).
46 H. Russell to H. Shapley, 17 September 1920, Harvard.
47 H. Shapley to H. Russell, 30 September 1920, Harvard.
48 H. Russell to H. Shapley, 12 October 1920, Harvard.
49 L. Silberstein to H. Shapley, 5 January 1924, Harvard.
50 L. Silberstein to H. Shapley, 17 April 1924, Harvard.
51 H. Shapley to A. Eddington, 31 May 1924, Harvard.
52 North points out that Silberstein's derivation was analytically correct (North (1965) 101–4).
53 Lundmark (1924b) 767.
54 Strömberg (1925) 362.
55 Strömberg (1925) 358.
56 Lundmark (1925). In 1936 Hubble wrote that 'Wirtz, the leader in the field, made the first attempt, in 1924, to express the K term as a function of distance using apparent diameters and velocities of 42 nebulae. A plausible correlation appeared in the expected direction– velocities tended to increase as diameters diminished' (Hubble (1936) 110). Wirtz expressed his results as a linear relation between velocities and the logarithm of the distances (which were derived from the diameters). But Wirtz's results appeared to include the effects of a relationship between the velocity and the concentration of a nebula. Also, Wirtz did not explicitly express the K term as a function of distance.
57 Dose (1927).
58 Weyl's Principle actually consisted of assuming that the stars (galaxies) can be treated as a uniformly distributed set of free particles which, except for small peculiar motions, remain at rest with respect to the spatial co-ordinates being used, but the proper distances measured by

rigid measuring rods will change with time due to a dependence of the metric tensor on time. Thus, a spectral shift is to be expected as light travels from one particle to another (Tolman (1934) 356–8).

59 Weyl (1923) 230.

60 Lemaître (1927).

61 Robertson (1928) 845.

62 Robertson (1928) 836. North writes that 'Robertson is fairly typical of his contemporaries. Whilst he was aware of the logical difficulties in interpreting "time" and "distance" variables, he offered no criterion by which his own particular choice could be justified' (North (1965) 118). We are not concerned here with the 'correctness' of the mathematical derivations or physical interpretations; our central interest is the perception of these analyses by contemporary astronomers.

63 Hubble (1926a) 368. Hubble used the formulae for the radius, mass and volume of the Einstein universe that were given in Haas (1925) 372–4.

64 M. Humason, A.I.P. Interview with Bert Shapiro c. 1965. In the leading astronomy text-book of the time, published in 1927, Russell, Dugan and Stewart took the de Sitter effect seriously and wrote that whether the red shifts of the nebulae represent 'a real scattering of the nebulae away from this region where the sun happens to be is very doubtful. It may arise from some other cause. Certain forms of the generalised theory of relativity . . . indicate that very distant luminous bodies appear to have large velocities of recession' (Russell, Dugan & Stewart (1927) 850).

65 See Hubble (1936) 102–3.

66 On Humason see Bowen (1972).

67 Hubble (1929b) 173.

68 Hubble (1929b) 174. Hubble was well aware of the importance of this result. In May 1929 he told Shapley that with his investigation a new phase of astronomy was opening (E. Hubble to H. Shapley, 15 May 1929, Harvard).

69 Hubble (1929b) 173.

70 This was not its universal usage. Eddington used the term 'de Sitter effect' to imply merely the apparent slowing down of atomic vibrations at great distances from the observer (Eddington (1933) 49).

71 Hubble (1929c) 738.

72 Hetherington (1970) 139.

73 E. Hubble to W. de Sitter, 21 August 1930, Huntington Library.

74 Shapley (1929b) 566.

75 Shapley (1929b) 567.

76 Hubble had mistaken what are now identified as H II regions–bright clouds of ionised gas – for the brightest stars. Since these clouds are intrinsically more luminous than stars, it meant that the corresponding distances were underestimated. This was, however, only demonstrated convincingly much later and within the astronomy of the period Hubble's assertion was entirely reasonable (see Sandage (1958) 522).

77 H. Shapley to E. Hubble, 7 May 1929, Harvard.

78 E. Hubble to H. Shapley, 15 May 1929, Harvard.

79 *Ibid.*

80 H. Shapley to H. Russell, 22 May 1929, Harvard.

81 Shapley, 'Progress in extragalactic explorations', notes on a symposium

address to the American Philosophical Society, 26 April 1930, Harvard.

82 de Sitter (1930*b*) 169. Eddington had told de Sitter of Lemaître's study a few weeks before he completed this paper.

83 Hubble nevertheless wanted to ensure that de Sitter recognised that the formulation, testing and confirmation of the velocity–distance relation was a 'Mount Wilson contribution', and he insisted to de Sitter that 'our preliminary note in 1929 was the first presentation of the data where the scatter due to uncertainties in distances was small enough as compared to the range in distances, to establish the relation' (E. Hubble to W. de Sitter, 21 August 1930, Huntington Library). Hubble was also chagrined that de Sitter had assigned equal weights to his own and Lundmark's distance estimates. In addition, as part of the programme to test the velocity–distance relation, Humason had measured the radial velocities of a number of nebulae and the values secured had then been published in the Mount Wilson Annual Reports. De Sitter had then employed these radial velocities in his own derivation of the linear relation. 'We have always assumed', Hubble protested, 'that, where a preliminary note is published and a program is announced for testing the result in new regions, the first discussion of the new data is reserved as a matter of courtesy to those who do the actual work. Are we to infer that you do not subscribe to this ethics; that we must hoard our observations in secret? Surely there is a misunderstanding somewhere?' Their later correspondence was amicable, but de Sitter, as he revealed to Shapley, was surprised by this letter (H. Shapley to H. Russell, 26 November 1931, Harvard).

84 Tolman (1929*a*) 245; italics in original.

85 R.A.S. (1930) 38.

86 Friedmann (1922).

87 Tolman (1929*b*). In this paper Tolman wrote that 'it should be noted that our assumption of a static line element takes no explicit recognition of any universal evolutionary process which may be going on. The investigation of non-static line elements would be very interesting' (p.304).

88 R.A.S. (1930) 39.

89 G. C. McVittie, who in 1929 was a research student of Eddington's, recalled in 1967 'the day when Eddington, rather shamefacedly, showed me a letter from Lemaître which reminded Eddington of the solution to the problem which Lemaître had already given. Eddington confessed that, though he had seen Lemaître's paper in 1927, he had completely forgotten about it until that moment' (McVittie (1967) 295). Early in 1930 Eddington sent de Sitter a copy of Lemaître's 1927 paper. Across the top of the front page Eddington wrote: 'This seems a complete answer to the problem we were discussing' (Leiden). There seems little doubt that the 'problem' was to find a suitable alternative to Solutions A and B.
  On the reception of the idea of non-static solutions, also see Hetherington (1973).

90 de Sitter (1930*a*) 482.

91 W. de Sitter to H. Shapley, 17 April 1930, Harvard.

92 de Sitter (1931) 584.

93 *New York Times*, 3 January 1931, 1.

94 Einstein & de Sitter (1932).
95 North (1965) 111. See also 111–21.
96 Friedmann (1922). See also Friedmann (1924).
97 Einstein had at first rejected the Friedmann metric as the result of a mistaken derivation, but he later retracted this criticism: see Einstein (1922) and (1923). Professor John Stachel writes that 'examination of the manuscript of Einstein's retraction of his criticism of the Friedmann paper shows that it orginally ended with a line to the effect that of course the solution, while mathematically correct, was of no physical significance! Fortunately, it was crossed out in manuscript' (J. Stachel to author, private communication).
98 Whitrow (1978) 582.
99 This point was not immediately obvious, and Eddington even had to bring it to de Sitter's notice (A. Eddington to W. de Sitter, 25 August 1930, Leiden).
100 It is relevant here to enquire if Hubble was aware of the explicit predictions by Weyl, Lemaître and Robertson of a linear redshift–distance relation when he published his 1929 paper. We know that in his research programme Hubble had set himself the task of testing de Sitter's model. In 1929 he had also noted that attempts had been made to explain the presence of a K term as a 'correlation between apparent radial velocities and distances, but so far the results have not been convincing' (Hubble (1929*b*) 168). Hubble would have known of the papers of Silberstein, Strömberg and Lundmark and their attempts to secure observationally a redshift–distance relation. It is unthinkable that he had not read them: Strömberg was, after all, at the same Observatory as Hubble; Lundmark's papers on a redshift–distance relation had appeared in the highly influential and very widely read *Monthly Notices of the Royal Astronomical Society,* as had Silberstein's 'The curvature of de Sitter's space–time derived from globular clusters'. Yet it is not clear from his public pronouncements whether or not Hubble was acquainted with the theoretical studies of Weyl, Robertson and Lemaître. In his 1929 paper Hubble did not mention any mathematical researchers with whom he had been in contact. In addition, a letter in the Hubble papers from H. P. Robertson to Hubble, probably dated sometime in 1933, suggests that in 1929 Hubble was not aware of the mathematical predictions of a linear relation since Robertson listed the predictions for him. Against this evidence is Hubble's contact with R. C. Tolman. Tolman was a Professor of Physical Chemistry and Mathematical Physics at the California Institute of Technology and he was a leading exponent of applying general relativity to cosmology. In 1934 his classic textbook *Relativity, thermodynamics and cosmology* was published and during the 1930s he collaborated with Hubble in studies of galaxies. In 1929 Tolman wrote a paper 'On the astronomical implications of the de Sitter line element for the universe'. This paper is dated 25 February. Hubble's paper had been sent to the National Academy of Sciences on 17 January but did not appear in the Society's *Proceedings* until the 15 March issue. But Tolman had emphasised in his paper that one of the three well established facts about the galaxies was that they exhibited a relation between velocity and distance of the form $(d\lambda/\lambda) = \alpha(r/R)$, where α is a constant. Tolman termed this Hubble's relationship.

Clearly he must have known of Hubble's research before Hubble's paper, which we will recall was held for over a year before publication, was published in the *Proceedings*. How much Tolman knew is unclear: maybe Tolman and Hubble had discussed Hubble's research programme. Perhaps Tolman even told Hubble about the predictions of a linear redshift–distance relation. Such a conjecture is strengthened when we note that Hetherington has pointed out that Tolman was based at Pasadena and that the offices of the Mount Wilson Observatory were sited in Pasadena. Also, in 1928 both Hubble and Tolman had been appointed to an advisory committee that was involved in the planning of the 200-inch telescope that was to be financed by the Rockefeller Foundation. Thus, Hubble had contact with at least one theoretician interested in and aware of the latest developments in relativisitc cosmology. Under these circumstances it is quite possible that Hubble knew of the theoretical predictions of a linear relation prior to the publication of the 1929 paper.

P. J. E. Peebles has surmised that because Lemaître's 1927 value for K ($600 \text{ km s}^{-1} \text{Mpc}^{-1}$) was so close to Hubble's 1929 value ($500 \text{ km s}^{-1} \text{Mpc}^{-1}$) 'there must have been communication of some sort between the two' (Peebles (1971) 8). There is, however, no documentary evidence for this; certainly there are no letters in the Hubble papers which corroborate Peebles's guess, and if Peebles is correct, then the neglect of Lemaître's paper until 1930 is more surprising.

101 Hubble (1929*b*) 173.
102 This point has been emphasised by Sandage (see Sandage (1975) 764).
103 North (1965) 92 ff.
104 Hubble (1936) 4.
105 Hubble and Humason (1931). This paper is described in some detail in Hetherington (1970) 144–52.
106 Humason (1931) 37. When Humason had started these investigations he had to use a very slow spectrograph and he had initially been rather unsure about securing spectrograms that required such long exposures. However, Hale had encouraged Humason to continue and promised that he would try to get a better spectrograph (M. Humason, A.I.P. Interview with Bert Shapiro, *c.* 1965).
107 Shapley (1930*a*) 189.
108 By 1935 even the careful Hubble felt able to write that 'the velocity–distance relation is so firmly established that it is assumed to be true for all nebulae, and the observed residuals are analysed for the information they give concerning the scatter in the intrinsic luminosities or the luminosity function of nebulae' (Hubble (1936) 120). In fact, a linear velocity–distance relation is an immediate consequence of the assumed homogeneity and isotropy of the Universe. This was first pointed out in 1935 by E. A. Milne (Milne (1935) 73). See also Peebles (1971) 10.
109 In 1977 N. U. Mayall, who in 1931 was working at Mount Wilson, recalled that even some of Humason's own colleagues suspected for a time that he had not properly identified the spectral lines in the spectrograms and had thereby exaggerated the redshifts (N. U. Mayall, A.I.P. interview with B. Shapiro, 13 February 1977).
110 Jeans (1928) 409.

111 Eddington (1933) 68. The 'Hubble constant' is K in our notation.

112 North (1965) 228.

113 E. Hubble to H. Shapley, 15 May 1929, Harvard.

114 E. Hubble to W. de Sitter, 23 September 1931, Huntington Library. By 1931 de Sitter's model had dropped from favour and Hubble could not claim that the linear relation demonstrated the reality of the de Sitter effect.

115 Zwicky (1929*a, b*). See also Hetherington (1970) 166–71 and North (1965) 229–34.

116 North (1965) 229. On the debate on the time-scale problem to the 1940s see Bok (1946).

117 de Sitter (1932*b*). Eddington advocated the view that the expansion had started from the Einstein static case.

118 For example, in 1922, Friedmann had talked about the period between space being a point ($R = 0$) and the present state as the time since the creation of the Universe: Friedmann (1922).

119 Jeans (1929) 327.

Bibliography

Brief note on sources

The *Astronomischer Jahresbericht* with its yearly list of abstracts of published papers is an indispensable source of information for the student of early twentieth century astronomy. The biographies in the *Biographical Memoirs of the National Academy of Sciences* of four of the major figures in the island universe debate – Hubble, Shapley, Slipher and Curtis – include a complete list of their books and published papers, and a list of Jeans's works is included in Milne (1952). van Maanen's major papers on internal motions in spiral nebulae are listed in Hart (1973). The *Dictionary of Scientific Biography* is also a useful starting point for securing information on many of the leading astronomers active between 1900 and 1931.

Historians of early twentieth century astronomy are fortunate in having two important manuscript collections – the Hale papers and the 'Early correspondence of the Lowell Observatory 1894–1916' – readily available on microfilm. A listing of the location of the archive materials of numerous American astronomers is given in *Journal of the History of Astronomy*, **2**, (1971), 210–218. The A.I.P. has extended this listing, not just by locating the materials of more American astronomers, but also by including Europeans. The A.I.P. also has a remarkably extensive collection of oral history interviews with leading twentieth century astronomers.

All items referred to in the notes to the text are listed below. No attempt has been made to separate primary and secondary sources as the distinction in any case should be clear. The manuscript sources that have been used have been fully referenced in the notes to the text.

A.A. Cyclopaedia (1866) *The American annual cyclopaedia and register of important events of the year 1865*. New York: D. Appleton & Co.

Adams, W. (1923) Annual report of the Director of Mount Wilson, *Yearb. Carnegie Inst. Wash.*, No. **22**, 181–217.

Adams, W. (1925) Annual report of the Director of Mount Wilson, *Yearb. Carnegie Inst. Wash.*, No. **24**, 89–126.

Adams, W. (1931) Annual report of the Director of Mount Wilson, *Yearb. Carnegie Inst. Wash.*, No. **30**, 171–221.

Adams, W. (1935) Annual report of the Director of Mount Wilson, *Yearb. Carnegie Inst. Wash.*, No. **34**, 157–90.

Aitken, R. (1943) Heber Doust Curtis 1872–1942, *Biogr. Mem. Nat. Acad. Sci.*, **22**, 275–94.

Baade, W. (1963) *Evolution of stars and galaxies*. Cambridge, Mass: Harvard University Press.

Badash, L. (1972) The completeness of nineteenth century science, *Isis*, **63**, 48–58.

Bailey, S. (1922) Henrietta Swan Leavitt, *Pop. Astron.*, **30**, 197–9.

Ball, R. (1907) *In the high heavens*, Cheap edition. London: Pitman & Sons.

Berendzen, R. (1975) Geocentric to heliocentric to galactocentric to acentric: The continuing assault to the egocentric, *Vistas Astron.*, **17**, 65–83.

Berendzen, R. & Hart, R. (1971) Hubble's classification of non-galactic nebulae, 1922–1926, *J. Hist. Astron.*, **2**, 109–19.

Berendzen, R. & Hart, R. (1973) Adriaan van Maanen's influence on the island universe theory, *J. Hist. Astron.*, **4**, 46–56 & 73–98.

Berendzen, R. & Hoskin, M. (1971) Hubble's announcement of Cepheids in spiral nebulae, *Astron. Soc. Pac.*, Leaflet No. 504, June 1971.

Berendzen, R. & Shamieh, C. (1973) Maanen, Adriaan van, *Dict. Sci. Biogr.*, **8**, 582–3.

Berendzen, R., Hart, R. & Seeley, D. (1976) *Man discovers the galaxies*. New York: Science History Publications.

Berry, A. (1898) *A short history of astronomy*. London: J. Murray.

Bohlin, K. (1907) Versuch einer Bestimmung der Parallaxe des Andromedanebels, *Astron. Nachr.*, **176**, 205–6.

Bohlin, K. (1909) On the galactic system with regard to its structure, origin, and relations in space, *Küngl Svenska Vetensamps Akadamiens Handlingar.*, **43**, No. 10.

Blackmore, J. (1972) *Ernst Mach: his life, work, and influence*. London: University of California Press.

Bohr, N., Kramers, H. & Slater, J. (1924) The quantum theory of radiation, *Zeitschr. Phys.*, **24**, 69–87.

Bok, B. (1946) The time-scale of the Universe, *Mon. Notic. Roy. Astron. Soc.*, **106**, 61–75.

Bok, B. (1978) Harlow Shapley, *Biogr. Mem. Nat. Acad. Sci.*, **49**, 240–91.

Bowen, I. (1972) Milton Lasel Humason, *Quart. J. Roy. Astron. Soc.*, **14**, 235–6.

Brush, S. (1978) A geologist among astronomers: The rise and fall of the Chamberlin–Moulton cosmogony, *J. Hist. Astron.*, **9**, 1–41 & 77–104.

Burchfield, J. (1975) *Lord Kelvin and the age of the Earth*. London: Macmillan.

Campbell, W. (1911) Some peculiarities in the motions of the stars, *Lick Observ. Bull.*, **6**, 124–35.

Campbell, W. (1913) *Stellar motions. With special reference to motions determined by means of the spectrograph*. New Haven: Yale University Press.

Campbell, W. (1917) The nebulae, *Science*, **45**, 513–48.

Campbell, W. (1926) Do we live in a spiral nebula?, *Publ. Astron. Soc. Pac.*, **38**, 75–85.

Charlier, C. (1925) An infinite universe, *Publ. Astron. Soc. Pac.*, **37**, 177–91.

Clark, R. (1973) *Einstein: the life and times*. London: Hodder & Stoughton.

Clerke, A. (1885) *A popular history of astronomy during the nineteenth century*. Edinburgh: Black.

Clerke, A. (1890) *The system of the stars*, First Edition. London: Longmans.

Clerke, A. (1903) *Problems in astrophysics*. London: Black.

Clerke, A. (1905) *The system of the stars*, Second Edition. London: Black.

Comstock, G. (1901) *A text-book of astronomy*. New York: Appleton & Co.

Crommelin, A. (1912) Notes: Astronomy, *Knowledge*, **9**, 31–2.

Crommelin, A. (1917) Are the spiral nebulae external galaxies?, *Scientia*, **21**, 365–76.

Curtis, H. (1911) The distances of the stars, *Publ. Astron. Soc. Pac.*, **23**, 143–63.

Curtis, H. (1915a). Preliminary note on nebular proper motions, *Proc. Nat. Acad. Sci.*, **1**, 10–12.

Curtis, H. (1915b) Proper motions of the nebulae, *Publ. Astron. Soc. Pac.*, **27**, 214–8.

Curtis, H. (1917a) The nebulae, *Publ. Astron. Soc. Pac.*, **29**, 91–103.

Curtis, H. (1917b) New stars in spiral nebulae, *Publ. Astron. Soc. Pac.*, **29**, 180–2.
Curtis, H. (1917c) Three novae in spiral nebulae, *Lick Observ. Bull.*, **9**, 108–10.
Curtis, H. (1921) The scale of the universe, *Bull. Nat. Res. Counc.*, **2**, part 3, 194–217.
Curtis, H. (1924) The spiral nebulae and the constitution of the universe, *Scientia*, **35**, 1–9.
Davidson, C. (1923) Astronomical photography, in *Photography as a scientific instrument* (London), a collective work, 209–61.
Dose, A. (1927) Zur Statistik der nichtgalaktischen Nebel auf Grund der Köngistuhl-Nebellisten: Mit einer Bewerkung über der Radialbewegungen der Spiralnebel, *Astron. Nachr.*, **229**, 157–76.
Douglas, A. (1956) *The Life of Arthur Stanley Eddington*. London: Nelson.
Duncan, J. (1922) Three variable stars and a suspected nova in the spiral nebula M 33 Trianguli, *Publ. Astron. Soc. Pac.*, **34**, 290–1.
Dyson, F. (1910) *Astronomy: A handy manual for students and others*. London: J. M. Dent & Sons.
Easton, C. (1900) A new theory of the Milky Way, *Astrophys. J.*, **12**, 136–58.
Easton, C. (1913) A photographic chart of the Milky Way and the spiral structure of the galactic system, *Astrophys. J.*, **37**, 105–18.
Eddington, A. (1912) Stellar distribution and movements, in *Report of the eighty-first meeting of the British Association for the Advancement of Science*. London: John Murray, 246–60.
Eddington, A. (1914) *Stellar movements and the structure of the universe*. London: Macmillan.
Eddington, A. (1915) The dynamics of a globular stellar system, *Mon. Notic. Roy. Astron. Soc.*, **75**, 366–76.
Eddington, A. (1916a) The nature of globular clusters, *The Observatory*, **39**, 513–4.
Eddington, A. (1916b) Internal motions in a spiral nebula, *The Observatory*, **39**, 514–5.
Eddington, A. (1917a) The motions of spiral nebulae, *Mon. Notic. Roy. Astron. Soc.*, **77**, 375–7.
Eddington, A. (1917b) Researches on globular clusters, *The Observatory*, **40**, 394–401.
Eddington, A. (1923) *The mathematical theory of relativity*. Cambridge University Press.
Eddington, A. (1930) *The rotation of the Galaxy*. Oxford: Clarendon Press
Eddington A. (1933) *The expanding universe*. Cambridge University Press.
Eddington, A. (1938) Forty years of astronomy, in *The background to modern science*, edited by J. Needham & W. Pagel, 115–42. Cambridge University Press.
Einstein, A. (1917) Kosmologische Betrachtungen zur allgemeinen Relativitätstheorie, *Sitzungsber. Königlich Preussischen Akad. Wiss.*, 142–52. This is printed in translation in *The principle of relativity* (Dover edition, 1952).
Einstein, A. (1919) Spielen Gravitationsfelder im Aufber der Materiellen Elementarteilchen eine wesentliche Rolle?, *Sitzungsber. Königlich Preussischen Akad. Wiss.*, 349–56. This is printed in translation in *The principle of relativity*, 191–8: see Einstein (1917).
Einstein, A. (1922) Bemerkung zu der Arbeit von A. Friedmann 'Über die Krummung des Raumes', *Zeitschr. Phys*, **11**, 326.
Einstein, A. (1923) Notiz zu der Arbeit von A. Friedmann 'Über die Krummung des Raumes', *Zeitschr. Phys.*, **16**, 228.
Einstein, A. & de Sitter, W. (1932) On the relation between the expansion and the mean density of the universe, *Proc. Nat. Acad. Sci.*, **18**, 213–4.
Fath, E. (1909) The spectra of some spiral nebulae and globular star clusters, *Lick Observ. Bull.*, **5**, 71–7.
Fath, E. (1910) The distribution of nebulae and globular star clusters, *Pop. Astron.*, **18**, 544–8.

Fath, E. (1911) The spectra of spiral nebulae and globular star clusters, second paper, *Astrophys. J.*, **33**, 58–63.

Fath, E. (1912) The story of the spirals, *Century Magazine*, **62**, 757–67.

Fath, E. (1913) The spectra of spiral nebulae and globular star clusters, third paper, *Astrophys. J.*, **37**, 198–203.

Fath, E. (1914) A study of nebulae, *Astron. J.*, **28**, 75–86.

Fath, E. (1926) *The elements of astronomy*. New York: McGraw-Hill.

Fernie, J. (1969) The period–luminosity relation: A historical review, *Publ. Astron. Soc. Pac.*, **81**, 707–31.

Fernie, J. (1970) The historical quest for the nature of the spiral nebulae, *Publ. Astron. Soc. Pac.*, **82**, 1189–1230.

Forbes, G. (1916) *David Gill, man and astronomer*. London: John Murray.

Fotheringham, J. (1926) Precession, galactic rotation, and equinox correction, *Mon. Notic. Roy. Astron. Soc.*, **86**, 414–26.

Friedmann, A. (1922) Über die Krümmung des Raumes, *Zeitschr. Phys.*, **10**, 377–86.

Friedmann, A. (1924) Über die Möglichkeit einer Welt mit Konstanter negativer Krümmung des Raumes, *Zeitschr. Phys.*, **21**, 326–32.

Frost, E. (1933) *An astronomer's life*. Boston, U.S.A: Houghton Mifflin.

Gaposchkin, S. & Payne-Gaposchkin, C. (1966) Variable stars in the Small Magellanic Cloud, *Smithsonian Contrib. Astrophys.*, **9**, 1–205.

Gingerich, O. (1973) Harlow Shapley and Mount Wilson, *Amer. Acad. Arts Sci. Bull.*, **26**, 10–24.

Gingerich, O. (1975a) Shapley, Harlow, *Dict. Sci. Biogr.*, **12**, 345–52.

Gingerich, O. (1975b) Dissertatio cum Professore Righini et Sidereo Nuncio, in *Reason, experiment and mysticism in the scientific revolution* (New York), edited by M. Bonelli and W. Shea, 77–88.

Gore, J. (1898) The sidereal heavens, in *The concise knowledge astronomy* edited by A. Clerke, J. Gore & A. Fowler, 399–565. London: Hutchinson & Co.

Gould, S. (1978) Morton's ranking of races by cranial capacity, *Science*, **200**, 503–9.

Graff, K. (1922) *Astrophysik: Dritte Völlig Neuarbeitere Auflage von. J. Scheiner Populäre Astrophysik*. Berlin: Teubner.

Haarh, G. & Luplau-Janssen, C. (1922) Die Parallaxe des Andromedanebels, *Astron. Nachr.*, **215**, 285.

Haas, A. (1925) *Introduction to theoretical physics*, Volume 2, New York: van Nostrand.

Hale, G. (1908) *The study of stellar evolution: An account of some recent methods of astrophysical research*. London: University of Chicago.

Hale, G. (1916) Annual report of the Director of Mount Wilson, *Yearb. Carnegie Inst. Wash.*, No. **15**, 227–72.

Hale, G. (1922) *The new heavens*. New York: Scribner & Sons.

Hall, A. (1870) On the secular perturbations of the planets, *Amer. J. Sci. & Arts*, **50**, Second series, 370–2.

Hall, J. (1970) Vesto Melvin Slipher, 1875–1969, *Yearb. Amer. Phil. Soc.*, 161–6.

Hall, M & van Tassell, D (1966) eds. *Science and society in the United States*. Illinois: Dorsey Press.

Hart, R. (1973) *Adriaan van Maanen's influence on the island universe theory*, unpublished Ph.D. thesis, Boston University Graduate School.

Herschel, J. (1864) Catalogue of nebulae and clusters of stars, *Phil. Trans. Roy. Soc.*, **154**, 1–137.

Hertzsprung, E. (1913) Über die räumliche Verteilung der Veränderlichen vom δ-Cephei-Typus, *Astron. Nachr.*, **196**, 201–10.

Hertzsprung, E. (1924) Report on the principal lines of research followed by me at the

Union Observatory, Johannesburg; from 1923 November to 1924 October, *Proc. Roy. Acad. Sci. Amsterdam*, **27**, 892–6.

Hetherington, N. (1970) *The development and early application of the velocity–distance relation*, unpublished Ph.D. thesis, Indiana University.

Hetherington, N. (1971) The measurement of radial velocities of spiral nebulae, *Isis*, **62**, 309–13.

Hetherington, N. (1972) Adriaan van Maanen and internal motions in spiral nebulae: A historical review, *Quart. J. Roy. Astron. Soc.*, **13**, 25–39.

Hetherington, N. (1973) The delayed response to suggestions of an expanding universe, *J. Brit. Astron. Ass.*, **84**, 22–8.

Hetherington, N. (1974a) Adriaan van Maanen on the significance of internal motions in spiral nebulae, *J. Hist. Astron.*, **5**, 52–3.

Hetherington, N. (1974b) Edwin Hubble's examination of internal motions of spiral nebulae, *Quart. J. Roy. Astron. Soc.*, **15**, 392–418.

Hetherington, N. (1975a) The simultaneous 'discovery' of internal motions in spiral nebulae, *J. Hist. Astron.*, **6**, 115–25.

Hetherington, N. (1975b) Adriaan van Maanen's measurements of solar spectra for a general magnetic field, *Quart. J. Roy. Astron. Soc.*, **16**, 235–44.

Holton, G. (1973) *Thematic origins of scientific thought: Kepler to Einstein*. Cambridge, Mass: Harvard University Press.

Hoskin, M. (1967) Apparatus and ideas in mid-nineteenth century cosmology, *Vistas Astron.*, **9**, 79–85.

Hoskin, M. (1968) Edwin Hubble and the existence of external galaxies, *Actes du XII Congrès International d'Histoire des Sciences*, 49–53. Paris: Albert Blanchard.

Hoskin, M. (1973) Dark skies and fixed stars, *J. Brit. Astron. Ass.*, **83**, 254–62.

Hoskin, M. (1976a) Ritchey, Curtis and the discovery of novae in spiral nebulae, *J. Hist. Astron.*, **7**, 47–53.

Hoskin, M. (1976b) The 'Great Debate': What really happened, *J. Hist. Astron.*, **7**, 169–82.

Hoyt, W. (1976) *Lowell and Mars*. Tucson: University of Arizona Press.

Hoyt, W. (1980) Vesto Melvin Slipher: 1875–1969, *Biogr. Mem. Nat. Acad. Sci.*, **52**, 411–49.

Hubble, E. (1920) Photographic investigations of giant nebulae, *Publ. Yerkes Observ.*, **4**, 69–85.

Hubble, E. (1922a) A general study of the diffuse galactic nebulae, *Astrophys. J.*, **56**, 162–99.

Hubble, E. (1922b) The source of luminosity in galactic nebulae, *Astrophys. J.*, **56**, 400–38.

Hubble, E. (1923) Messier 87 and Belanowsky's novae, *Publ. Astron. Soc. Pac.*, **35**, 261–3.

Hubble, E. (1925) NGC 6822, a remote stellar system, *Astrophys. J.*, **62**, 409–33.

Hubble, E. (1926a) A spiral nebula as a stellar system, Messier 33, *Astrophys. J.*, **63**, 236–74.

Hubble, E. (1926b) Extra-galactic nebulae, *Astrophys. J.*, **64**, 321–69.

Hubble, E. (1929a) A spiral nebula as a stellar system, Messier 31, *Astrophys. J.*, **69**, 103–58.

Hubble, E. (1929b) A relation between distance and radial velocity among extra-galactic nebulae, *Proc. Nat. Acad. Sci.*, **15**, 168–73.

Hubble, E. (1929c) The exploration of space, *Harper's Magazine*, **58**, 732–8.

Hubble, E. & Humason, M. (1931) The velocity-distance relation among extra-galactic nebulae, *Astrophys. J.*, **74**, 43–80.

Hubble, E. (1935) Angular rotations of spiral nebulae, *Astrophys. J.*, **81**, 334–5.

Hubble, E. (1936) *The realm of the nebulae*. Oxford University Press.

Huggins, W. (1865) From an account of his researches on the prismatic analysis of nebulae given to the Royal Astronomical Society on 10 March 1865, *Mon. Notic. Roy. Astron. Soc.*, **25**, 155–7.

Huggins, W. (1868) Further observations on the spectra of some of the stars and nebulae, with an attempt to determine therefrom whether these bodies are moving towards or from the Earth also observations on the spectra of the sun and of Comet II, 1868, *Phil. Trans. Roy. Soc.*, **158**, 529–64.

Huggins, W. & Lady Huggins (1899) *An atlas of representative stellar spectra from λ 4870 to λ 3300*. London: Wesley & Son.

Huggins, W. & Lady Huggins (1909) *The scientific papers of Sir William Huggins*. London: Wesley & Son.

Humason, M. (1929) The large radial velocity of NGC 7619, *Proc. Nat. Acad. Sci.*, **15**, 167–8.

Humason, M. (1931) Apparent velocity-shifts in the spectra of faint nebulae, *Astrophys. J.*, **74**, 35–42.

Humboldt, A. Von (1850) *Kosmos: Entwurf einer physischen Weltbeschreibung*, volume 3. Stuttgart and Tubingen: Cotta. Translated by E. Otté as *Cosmos: A sketch of a physical description of the universe*, volume 3. London, 1851:Bohn.

Huxley, T. (1869) Review of the present state of geological speculation, *Quart. J. Geol. Soc. London*, **25**, xxxviii–liii.

I.A.U. (1925) *Transactions of the International Astronomical Union*, **2**.

I.A.U. (1928) *Transactions of the International Astronomical Union*, **3**.

I.A.U. (1932) *Transactions of the International Astronomical Union*, **4**.

Jackson, J. (1924) On the influence of comparison stars on photographic proper motions, *Mon. Notic. Roy. Astron. Soc.*, **84**, 401–9.

Jaki, S. (1969) *The paradox of Olbers' paradox: A case history of scientific thought*. New York: Herder & Herder.

Jaki, S. (1973) *The Milky Way: An elusive road for science*. Newton Abbot: David & Charles.

Jaki, S. (1978) *Planets and Planetarians: A history of the origin of planetary systems*. Edinburgh: Scottish Academic Press.

Jammer, M. (1966) *The conceptual development of quantum mechanics*. New York: McGraw-Hill.

Jeans, J. (1917) Internal motion in spiral nebulae, *The Observatory*, **40**, 60–1.

Jeans, J. (1919) *Problems of cosmogony and stellar dynamics*. Cambridge University Press.

Jeans, J. (1921a) Address by Jeans at the Royal Astronomical Society meeting of November 1921, *The Observatory*, **44**, 352–5.

Jeans, J. (1921b) Cosmogony and stellar evolution, *Nature*, **107**, 557–60 & 588–90.

Jeans, J. (1923a) Internal motions in spiral nebulae, *Mon. Notic. Roy. Astron. Soc.*, **84**, 60–76.

Jeans, J. (1923b) Review of Eddington's *The mathematical theory of relativity*, *The Observatory*, **46**, 191–3.

Jeans, J. (1924) The origin of the solar system, *Notic. Proc. Roy. Inst. Great Britain*, **24**, 313–32.

Jeans, J. (1925) Note on the distances and structure of the spiral nebulae, *Mon. Notic. Roy. Astron. Soc.*, **85**, 531–4.

Jeans, J. (1928) *Astronomy and cosmogony*. Cambridge University Press.

Jeans, J. (1929) *The universe around us*. Cambridge University Press.

Johnson, M. (1950) John Henry Reynolds, *Mon. Notic. Roy. Astron. Soc.*, **110**, 131–3.

Johnson, M. (1961) *Knut Lundmark och världsrymdens erövring. En minnesskrift* [Knut Lundmark and man's march into space. A memorial volume] (Göteborg).

Jones, H. (1915) On the structure of the universe, *Science Progress*, **10**, 1–16.

Jones, H. (1922) *General astronomy*. London : Arnold & Co.

Jones, K. (1971) The observational basis for Kant's *Cosmogony*: A critical analysis, *J. Hist. Astron.*, **2**, 29–34.

Jones, K. (1976) S Andromedae, 1885: An analysis of contemporary reports and a reconstruction, *J. Hist. Astron.*, **7**, 27–40.

Kahn, F. & Kahn, C. (1975) Letters from Einstein to de Sitter on the nature of the universe, *Nature*, **257**, 451–4.

Kapteyn, J. (1908) Recent researches in the structure of the universe [from an address at the Royal Institution in 1908] in *The Royal Institution Library of Science, Astronomy*, volume 2, edited by B. Lovell, 78–96. Barking, 1970: Elsevier Publishing Co.

Kapteyn, J. (1913) On the structure of the universe, *Scientia*, **14**, 345–57.

Kapteyn, J. (1914) On the individual parallaxes of the brighter galactic helium stars in the Southern Hemisphere, together with considerations on the parallax of stars in general, *Astrophys. J.*, **40**, 43–126.

Kapteyn, J. (1922) First attempt at a theory of the arrangement and motion of the sidereal system, *Astrophys. J.*, **55**, 302–28.

Kapteyn, J & van Rhijn, P. (1920) On the distribution of the stars in space especially in the high galactic latitudes, *Astrophys. J.*, **52**, 23–38.

Kapteyn, J. & van Rhijn, P. (1922) The proper motions of δ-Celphei stars and the distances of the globular clusters, *Bull. Astron. Inst. Neth.*, **1**, 37–42.

Keeler, J. (1894) Spectroscopic observations of nebulae, *Lick Publ.*, **3**, 161–229.

Keeler, J. (1900) The Crossley Reflector of the Lick Observatory, *Astrophys. J.*, **11**, 325–349.

Kostinsky, S. (1917) Probable motions in the spiral nebula Messier 51 (Canes Venatici) found with the stereo-comparator, preliminary communication, *Mon. Notic. Roy. Astron. Soc.*, **77**, 233–4.

Lampland, C. (1916) Preliminary measures of the spiral nebula NGC 5194 (M 51) and NGC 4254 (M 99) for proper motion and rotation, *Pop. Astron.*, **24**, 667–8.

Lampland, C. (1918) On the proper motion of the Virgo nebula NGC 4594, *Publ. Amer. Astron. Soc.*, **3**, 83–4.

Langley, S. (1884) The new astronomy, *Century Magazine*, **6**, 712–26.

Leavitt, H. (1908) 1777 variables in the Magellanic Clouds, *Harvard Coll. Observ. Annals*, **60**, 87–108.

Lemaître, G. (1927) Un univers homogène de masse constante et de rayon croissant, rendant compte de la vitesse radiale des nébuleuses extra-galactiques, *Annales de la Société Scientifique de Bruxelles*, **47**, 49–56. This was translated as 'A homogeneous universe of constant mass and increasing radius accounting for the radial velocity of extra-galactic nebulae', *Mon. Notic. Roy. Astron. Soc.*, **91**, 483–90.

Lindblad, B. (1925a) On the cause of star-streaming, *Astrophys. J.*, **62**, 191–7.

Lindblad, B. (1925b) Star-streaming and the structure of the stellar system, *Meddelanden Från Astron. Observ. Uppsala*, Series C, **1**, No. 3.

Lindblad, B. (1927) On the nature of spiral nebulae, *Mon. Notic. Roy. Astron. Soc.*, **87**, 420–6.

Lindemann, F. (1923) Note on the constitution of the spiral nebulae, *Mon. Notic. Roy. Astron. Soc.*, **83**, 354–9.

Lundmark, K. (1920) The relations of the globular clusters and spiral nebulae to the stellar system. An attempt to estimate their parallaxes, *Küngl Svenska Vetensamps Akademiens Handlingar*, Band **60**, No. 8.

Lundmark, K. (1921) The spiral nebula Messier 33, *Publ. Astron. Soc. Pac.*, **33**, 324–7.

Lundmark, K. (1922) On the motions of spirals, *Publ. Astron. Soc. Pac.*, **34**, 108–15.

Lundmark, K. (1924a) Der gegenwärtige Stand des Problems der Entfernung der Spiralnebel, *Vierteljahresschr. Astron. Ges.*, **59**, 218–22.

Lundmark, K. (1924*b*) The determination of the curvature of space–time in de Sitter's world, *Mon. Notic. Roy. Astron. Soc.*, **84**, 747–70.

Lundmark, K. (1925) The motions and distances of spiral nebulae, *Mon. Notic. Roy. Astron. Soc.*, **85**, 865–94.

Lundmark, K. (1926*a*) A preliminary classification of nebulae, *Meddelanden Från Astron. Observ. Uppsala*, Series C, **1**, No. 7.

Lundmark, K. (1926*b*) Double spiral nebulae and the law of the variation of the absolute dimensions of anagalactic nebulae, *Meddelanden Från Astron. Observ. Uppsala*, Series C, **1**, No. 8.

Lundmark, K. (1927) Studies of anagalactic nebulae, first paper, *Meddelanden Från Astron. Observ. Uppsala*, Series A, **1**, No. 7.

Lundmark, K. (1930) Are the globular clusters and the anagalactic nebulae related?, *Publ. Astron. Soc. Pac.*, **42**, 23–30.

Luyten, W. (1926) Island universes, *Harvard Coll. Observ. Reprint* No. 32, reprinted from *Natural History*, **26**, 386–91.

Maanen, A. van (1916*a*) Preliminary evidence of internal motion in the spiral nebula Messier 101, *Astrophys. J.*, **44**, 210–28.

Maanen, A. van (1916*b*) Preliminary evidence of internal motion in the spiral nebula Messier 101, *Proc. Nat. Acad. Sci.*, **2**, 386–90.

Maanen, A. van (1921*a*) Investigations on proper motion, fourth paper: Internal motion in the spiral nebula Messier 51, *Astrophys. J.*, **54**, 237–45.

Maanen, A. van (1921*b*) Investigations on proper motion, fifth paper: The internal motion in the spiral nebula Messier 81, *Astrophys. J.*, **54**, 347–56.

Maanen, A. van (1921*c*) Internal motion in the spiral nebula Messier 33, preliminary results, *Proc. Nat. Acad. Sci.*, **7**, 1–5.

Maanen, A. van (1922*a*) Investigations on proper motion, seventh paper: Internal motion in the spiral nebula NGC 2403, *Astrophys. J.*, **56**, 200–7.

Maanen, A. van (1922*b*) Investigations on proper motion, eighth paper: Internal motion in the spiral nebula M 94 = NGC 4736, *Astrophys. J.*, **56**, 208–16.

Maanen, A. van (1923*a*) Investigations on proper motion, ninth paper: Internal motion in the spiral nebula Messier 63, NGC 5055, *Astrophys. J.*, **57**, 49–56.

Maanen, A. van (1923*b*) Investigations on proper motion, tenth paper: Internal motion in the spiral nebula Messier 33, NGC 598, *Astrophys. J.*, **57**, 264–78.

Maanen, A. van (1925*a*) Investigations on proper motion, eleventh paper: The proper motion of Messier 13 and its internal motion, *Astrophys. J.*, **61**, 130–6.

Maanen, A. van (1925*b*) R.A.S. meeting, *Mon. Notic. Roy. Astron. Soc.*, **85**, 901–2.

Maanen, A. van (1927) Investigations on proper motion, twelfth paper: The proper motions and internal motions of Messier 2, 13, and 56, *Astrophys. J.*, **66**, 89–112.

Maanen, A. van (1930*a*) Investigations on proper motion, fifteenth paper: The proper motion of the spiral nebulae NGC 4051, *Contrib. Mount Wilson Solar Observ.*, No. 407.

Maanen, A. van (1930*b*) Investigations on proper motion, sixteenth paper: The proper motion of Messier 51, NGC 5194, *Contrib. Mount Wilson Solar Observ.*, No. 408.

Maanen, A. van (1935) Internal motions in spiral nebulae, *Astrophys. J.*, **81**, 336–7.

Mach, E. (1902) *The science of mechanics*. Chicago: The Open Court Publishing Co. This is a translation by T. J. McCormack of the second edition of Mach's *Die Mechanik in ihrer Entwickelung historischkritisch dargestellt*.

MacMillan, W. (1920) Review of Jeans's *Problems of cosmogony and stellar dynamics*, *Astrophys. J.*, **51**, 309–33.

Macpherson, H. (1905) *Astronomers of today and their work*. London and Edinburgh: Gall & Inglis.

Macpherson, H. (1916) Some problems of astronomy – XXII – The nature of spiral nebulae, *The Observatory*, **39**, 131–4.

Mayall, N. (1954) Edwin Hubble: Observational cosmologist, *Sky & Telescope,* **13,** 78–81.
Mayall, N. (1970) Edwin Powell Hubble, *Biogr. Mem. Nat. Acad. Sci.,* **41,** 175–214.
McCrea, W. (1979) Cosmology after Einstein, *New Scientist,* **81,** 756–60.
McLaughlin, D. (1922) The present position of the island universe theory of the spiral nebulae, *Pop. Astron.,* **30,** 286–95 & 327–39.
McMath, R. (1944) Heber Doust Curtis, *Astrophys. J.,* **99,** 245–8.
McVittie, G. (1967) Georges Lemaître, *Quart. J. Roy. Astron. Soc.,* **8,** 294–7.
Miller, H. (1966) Science and private agencies, in M. Hall (1966), 191–221.
Miller, H. (1970) *Dollars for research; science and its patrons in nineteenth-century America.* Seattle: University of Washington Press.
Milne, E. (1935) *Relativity, Gravitation and World Structure.* Oxford: Clarendon Press.
Milne, E. (1952) *Sir James Jeans, a biography.* Cambridge University Press.
Moszkowski, A. (1921) *Einstein the searcher: his work explained from dialogues,* translated by H. L. Brose. London: Methuen & Co.
Moulton, F. (1905) On the evolution of the Solar System, *Astrophys. J.,* **22,** 165–81.
Moulton, F. (1906) *An introduction to astronomy.* London: Macmillan
Newcomb, S. (1888) The place of astronomy among the sciences, *Sidereal Messenger,* **7,** 14–20 & 65–73.
Newcomb, S. (1906) *Sidelights on astronomy and kindred fields of popular science.* New York and London: Harper & Bros.
Nichol, J. (1848) *Thoughts on some important points relating to the system of the world.* Edinburgh: Johnston.
Nielsen, A. (1963) Contributions to the history of the Hertzsprung-Russell diagram, *Centaurus,* **9,** 219–53.
North, J. (1965) *The measure of the universe: A history of modern cosmology.* Oxford: Clarendon Press.
Oort, J. (1926) The stars of high velocity, *Publ. Kapteyn Astron. Lab. Groningen,* No. 40.
Oort, J. (1927a) Observational evidence confirming Lindblad's hypothesis of a rotation of the galactic system, *Bull. Astron. Inst. Neth.,* **3,** 275–82.
Oort, J. (1927b) Investigations concerning the rotational motion of the galactic system, together with new determinations of secular parallaxes, precession and motion on the equinox, *Bull. Astron. Inst. Neth.,* **4,** 79–89.
Oort, J. (1966) Bertil Lindblad, *Quart. J. Roy. Astron. Soc.,* **7,** 329–41.
Oort, J. (1972) The development of our insight into the structure of the Galaxy between 1920 and 1940, *Annals New York Acad. Sci.,* **198,** 255–66.
Öpik, E. (1922) An estimate of the distance of the Andromeda Nebula, *Astrophys. J.,* **55,** 406–10.
Osterbrock, D. (1976) The California–Wisconsin Axis in American astronomy, II, *Sky & Telescope,* **51,** 91–7.
Paddock, G. (1916) The relation of the system of stars to the spiral nebulae, *Publ. Astron. Soc. Pac.,* **28,** 109–15.
Pannekoek, A. (1961) *A history of astronomy.* London: Allen & Unwin.
Parsons, C. (1926) *The scientific papers of William Parsons, third Earl of Rosse 1800–1867,* London: Lund Humphries & Co.
Paul, E. (1976) *Seeliger, Kapteyn and the rise of statistical astronomy,* unpublished Ph.D. thesis, Indiana University.
Pease, F. (1916a) The rotation and radial velocity of the spiral nebula NGC 4594, *Proc. Nat. Acad. Sci.,* **2,** 517–21.
Pease, F. (1916b) The rotation and radial velocity of the spiral nebula NGC 4594, *Publ. Astron. Soc. Pac.,* **28,** 191.
Pease, F. (1917) Photographs of nebulae with the 60-inch reflector, 1911–1916, *Astrophys. J.,* **46,** 24–55.

Pease, F. (1920) Photographs of nebulae with the 60-inch reflector, 1917–1919, *Astrophys. J.*, **51**, 276–308.

Peebles, P. (1971) *Physical cosmology*. Princeton University Press.

Philips, T. (1918) Variable stars, *Mon. Notic. Roy. Astron. Soc.*, **78**, 306–9.

Pickering, E. (1895) A new star in Centaurus, *Harvard Coll. Observ. Circular*, No. 4, 20 December 1895.

Pickering, E. (1912) Periods of 25 variable stars in the Small Magellanic Cloud, *Harvard Coll. Observ. Circular*, No. 173, 3 March 1912.

Pickering, E. (1917) Report of telegram from Adams to Pickering about Ritchey's nova in NGC 6946, *Harvard Coll. Observ. Bulletin*, No. 641, 28 July 1917.

Plaskett, H. (1931) The nebulae outside the Galaxy, *J. Roy. Astron. Soc. Can.*, **25**, 328.

Plaskett, J. (1911) Some recent interesting developments in astronomy, *J. Roy. Astron. Soc. Can.*, **5**, 245–65.

Plaskett, J. (1928) The rotation of the Galaxy, *Mon. Notic. Roy. Astron. Soc.*, **88**, 395–403.

Plaskett, J. (1935) *The dimensions and structure of the Galaxy*. Oxford: Clarendon Press.

Proctor, R. (1869) Distribution of the nebulae, *Mon. Notic. Roy. Astron. Soc.*, **29**, 337–44.

Puiseux, P. (1913) The spiral nebulae, *Annual report of the board of Regents of the Smithsonian Institution, 1912*.

R.A.S. (1916*a*) Report of the R.A.S. meeting in May 1916, *The Observatory*, **39**, 243–56.

R.A.S. (1916*b*) Report of the R.A.S. meeting in November 1916, *The Observatory*, **39**, 479–93.

R.A.S. (1921) Report of the R.A.S. meeting in November 1921, *The Observatory*, **44**, 351–61.

R.A.S. (1930) Report of the R.A.S. meeting in January 1930, *The Observatory*, **53**, 33–44.

Ravetz, J. (1973) *Scientific knowledge and its social problems* (Penguin edition).

Reynolds, J. (1912) Preliminary observations of spiral nebulae in polarised light, *Mon. Notic. Roy. Astron. Soc.*, **72**, 553–5.

Reynolds, J. (1916) The nature of spiral nebulae, *The Observatory*, **39**, 174–5.

Reynolds, J. (1917) Motion in the spiral nebulae M 101 and NGC 4954, *The Observatory*, **40**, 131–2.

Reynolds, J. (1920*a*) Photometric measures of the nuclei of some typical spiral nebulae, *Mon. Notic. Roy. Astron. Soc.*, **80**, 746–53.

Reynolds, J. (1920*b*) The galactic distribution of the large spiral nebulae, *Mon. Notic. Roy. Astron. Soc.*, **81**, 129–31.

Reynolds, J. (1922) The planes of the spiral nebulae in relation to the line of sight, *Mon. Notic. Roy. Astron. Soc.*, **82**, 510–13.

Reynolds, J. (1923) Star clusters and nebulae, in *Hutchinson's splendour of the heavens: A popular authoritative astronomy*, edited by T. Philips & W. Steavenson, 523–84. London: Hutchinson.

Reynolds, J. (1924) The condensations in the spiral nebulae, *Mon. Notic. Roy. Astron. Soc.*, **85**, 142–7.

Reynolds, J. (1926) Nebulae, *Mon. Notic. Roy. Astron. Soc.*, **86**, 257–9.

Ritchey, G. (1910) On some methods and results in direct photography with the 60-inch reflecting telescope of the Mount Wilson Observatory, *Astrophys. J.*, **32**, 26–35.

Roberts, I. (1888) Photographs of the nebulae M 31, h 44, and h 51 Andromedae, and M 27 Vulpeculae, *Mon. Notic. Roy. Astron. Soc.*, **49**, 65.

Robertson, H. (1928) On relativistic cosmology, *Philosophical Magazine*, series VII, **5**, 835–48.

Rothenberg, M. (1974) *The educational and intellectual background of American astronomers, 1825–1875*, unpublished Ph.D. thesis, Bryn Mawr College.

Russell, H. (1913) Notes on the real brightness of variable stars, *Science*, **37**, 651–2,

Russell, H. (1918) Astronomy notes, *The Scientific American*, **118**, 412.

Russell, H. (1919) Some problems of sidereal astronomy, *Proc. Nat. Acad. Sci.*, 5, 391–416.

Russell, H. & Shapley, H. (1914) On the distribution of eclipsing variable stars in space, *Astrophys. J.*, 40, 417–34.

Russell, H., Dugan, R. & Stewart, J. (1927) *Astrophysics and stellar astronomy, II.* Boston, U.S.A.: Ginn & Co.

Sandage, A. (1958) Current problems in the extragalactic distance scale, *Astrophys. J.*, 127, 513–26.

Sandage, A. (1975) The redshift, in *Galaxies and the Universe*, eds. Sandage, A., Sandage, M. & Kristian, J. University of Chicago.

Sandage, A. & Tammann, G. (1974) Steps toward the Hubble Constant. III. The distance and stellar content of the M 101 group of galaxies, *Astrophys. J.*, 194, 223–43.

Sanford, R. (1917) On some relations of the spiral nebulae to the Milky Way, *Lick Observ. Bull.*, 9, 80–91.

Scheiner, J. (1890) *Die Spectralanalyse der Gestirne.* Leipzig: Engelmann. This was translated by E. B. Frost (1894) as *A treatise on astronomical spectroscopy being a translation of Die Spectralanalyse der Gestirne by Prof. Scheiner* Boston, U.S.A.: Ginn & Co.

Scheiner, J. (1899) On the spectrum of the great Nebula in Andromeda, *Astrophys. J.*, 9, 149–50.

Schilt, J. (1926) Remarks on various statistical properties of galactic Cepheids having periods longer than one day, *Astrophys. J.*, 64, 149–66.

Schouten, W. (1919a) The parallax of some stellar clusters, *The Observatory*, 42, 112–9.

Schouten, W. (1919b) Probable motions in the spiral nebula Messier 51, *The Observatory*, 42, 441–4.

Schwarzschild, K. (1900) Ueber das zulässige Krümmungsmaass des Raumes, *Vierteljahresschr. Astron. Ges.*, 35, 337–47.

Seares, F. (1920) The surface brightness of the galactic system as seen from a distant external point and a comparison with a spiral nebula, *Astrophys. J.*, 52, 162–82.

Seares, F. (1927) Some structural features of the galactic system, *Astrophys. J.*, 67, 123–78.

Seares, F. (1946) Adriaan van Maanen, 1884–1946, *Publ. Astron. Soc. Pac.*, 58, 89–103.

Seeley, D. (1973) *The development of research on the interstellar medium c. 1900–1940: Diffuse nebulae, interstellar gas, and interstellar extinction*, unpublished Ph.D. thesis, Boston University Graduate School.

Seelinger, H. von (1898) Betrachtungen über die räumliche Vertheilung der Fixsterne, *Abhandlungen Math.-Phys. Classe Königlich Bayerischen Akad. Wiss.*, 19, 565–629.

Shapley, H. (1914) On the nature and cause of Cepheid variation, *Astrophys. J.*, 40, 448–65.

Shapley, H. (1915a) Studies based on the colors and magnitudes in stellar clusters. First paper: The general problem of clusters, *Contrib. Mount Wilson Solar Observ.*, No. 115.

Shapley, H. (1915b) Studies based on the colors and magnitudes in stellar clusters. Second paper: Thirteen hundred stars in the Hercules Cluster (Messier 13), *Contrib. Mount Wilson Solar Observ.*, No. 116.

Shapley, H. (1916) A short period Cepheid with variable spectrum, *Proc. Nat. Acad. Sci.*, 2, 132–6.

Shapley, H. (1917) Note on the magnitudes of novae in spiral nebulae, *Publ. Astron. Soc. Pac.*, 29, 213–17.

Shapley, H. (1918a) Globular clusters and the structure of the galactic system, *Publ. Astron. Soc. Pac.*, 30, 42–54.

Shapley, H. (1918b) Studies based on the colors and magnitudes in stellar clusters. Sixth paper: On the determination of the distances of globular clusters, *Astrophys. J.*, 48, 89–124.

Shapley, H. (1919*a*) Studies based on the colors and magnitudes in stellar clusters. Twelfth paper: Remarks on the arrangement of the sidereal universe, *Astrophys. J.*, **49**, 311–36.

Shapley, H. (1919*b*) On the existence of external galaxies, *Publ. Astron. Soc. Pac.*, **31**, 261–8.

Shapley, H. (1920) On the dimensions and arrangement of the galactic system [being part III of 'Star clusters and the structure of the universe'], *Scientia*, **27** (1920), 93–101.

Shapley, H. (1921) The scale of the universe, *Bull. Nat. Res. Counc.*, **2**, part 2, 171–193.

Shapley, H. (1922) Note on the problem of great stellar distances, *Proc. Nat. Acad. Sci.*, **8**, 69–71.

Shapley, H. (1923*a*) The galactic system, *Pop. Astron.*, **31**, 316–28.

Shapley, H. (1923*b*) Note on the distance of NGC 6822, *Harvard Coll. Observ. Bull.*, No. 796, 21 December 1923.

Shapley, H. (1923*c*) Note on Jeans' theory of the origin of stars, *Harvard Coll. Observ. Circular*, No. 257, 14 June 1923.

Shapley, H. (1927) On the classification of extra-galactic nebulae, *Harvard Coll. Observ. Bull.*, No. 849, 1 August 1927.

Shapley, H. (1929*a*) Studies of the galactic centre. IV. On the transparency of the galactic star clouds, *Proc. Nat. Acad. Sci.*, **15**, 174–7.

Shapley, H. (1929*b*) Note on the velocities and magnitudes of external galaxies, *Proc. Nat. Acad. Sci.*, **15**, 565–70.

Shapley, H. (1930*a*) *Star clusters*. New York: McGraw Hill.

Shapley, H. (1930*b*) *Flights from chaos: A survey of material systems from atoms to galaxies*. New York: McGraw Hill.

Shapley, H. (1930*c*) The super-galaxy hypothesis, *Harvard Coll. Observ. Circular*, No. 350, February 1930.

Shapley, H. (1969) *Through rugged ways to the stars*. New York: Scribner & Sons.

Shapley, H. & Ames, A. (1929) The Coma–Virgo galaxies, 1. On the transparency of inter-galactic space, *Harvard Coll. Observ. Bull.*, No. 864, 1 February 1929.

Shapley, H. & Shapley, M. (1919) Studies based on the colors and magnitudes in stellar clusters. Fourteenth paper: Further remarks on the structure of the galactic system, *Astrophys. J.*, **50**, 107–140.

Silberstein, L. (1924) The curvature of de Sitter's space–time derived from globular clusters, *Mon. Notic. Roy. Astron. Soc.*, **84**, 363–6.

Sitter, W. de (1917) On Einstein's theory of gravitation, and its astronomical consequences, third paper, *Mon. Notic. Roy. Astron. Soc.*, **78**, 3–28.

Sitter, W. de (1922) On the possibility of statistical equilibrium of the universe, *Proc. Roy. Acad. Sci. Amsterdam*, **22**, 866–8.

Sitter, W. de (1930*a*) On the distances and radial velocities of extra-galactic nebulae, and the explanation of the latter by the relativity theory of inertia, *Proc. Nat. Acad. Sci.*, **16**, 474–88.

Sitter, W. de (1930*b*) On the magnitudes, diameters and distances of the extragalactic nebulae, and their apparent radial velocities, *Bull. Astron. Inst. Neth.*, **5**, 157–71.

Sitter, W. de (1931) [Contribution to] Discussion on the evolution of the universe, *Report of the British Association for the Advancement of Science – London*, 583–7.

Sitter, W. de (1932*a*) *Kosmos*. Cambridge, Mass: Harvard University Press.

Sitter, W. de (1932*b*) On the expanding universe, *Proc. Roy. Acad. Sci. Amsterdam*, **35**, 596–607.

Slipher, V. (1912) On the spectrum of the nebula in the Pleiades, *Lowell Observ. Bull.*, No. 55.

Slipher, V. (1913) The radial velocity of the Andromeda Nebula, *Lowell Observ. Bull.*, No. 58.

Slipher, V. (1915) Spectrographic observations of nebulae, *Pop. Astron.*, **23**, 21–4.

214   Bibliography

Slipher, V. (1917) Nebulae, *Proc. Amer. Phil. Soc.*, **56**, 403–9.

Slipher, V. (1921) Dreyer nebula no. 584 inconceivably distant [an article on Slipher's measurement of the radial velocity of NGC 584], *New York Times*, 19 January 1921.

Smart, W. (1924) The motions of spiral nebulae, *Mon. Notic Roy. Sc.*, **84**, 333–53.

Smart, W. (1928) *The sun, the stars and the universe*. London: Longmans.

Smart, W. (1962) *Textbook on spherical astronomy*, Fifth edition Cambridge University Press.

Smith, R. (1977a) Russell and stellar evolution, [see de Vorkin & Philip (1977), 9–13].

Smith, R. (1977b) Sir James Hopwood Jeans, 1877–1946, *J. Brit. Astron. Ass.*, **88**. 8–17.

Stenflo, J. (1970) Hale's attempts to determine the sun's general magnetic field, *Solar Physics*, **14**, 263–70.

Stewart, L. (1913) The structure of the universe, *J. Roy. Astron. Soc. Can.*, **7**, 1–18.

Strömberg, G. (1925) Analysis of radial velocities of globular clusters and non-galactic nebulae, *Astrophys. J.*, **61**, 353–62.

Struve, O. (1857) Observations of the stars in the Nebula of Orion, *Mon. Notic. Roy. Astron. Soc.*, **17**, 225–30.

Struve, O. & Zebergs, V. (1962) *Astronomy of the twentieth century*, New York: Macmillan

Swihart, T. (1968) *Astrophysics and stellar astronomy*. New York: Wiley.

Todd, D. (1897) *A new astronomy* London: Sampson Low & Co.

Tolman, R. (1929a) On the astronomical implications of the de Sitter line element for the universe, *Astrophys. J.*, **69**, 245–74.

Tolman, R. (1929b) On the possible line elements for the universe, *Proc. Nat. Acad. Sci.*, **15**, 297–304.

Tolman, R. (1934) *Relativity, thermodynamics and cosmology*, Oxford : Clarendon Press.

Truman, O. (1916) The motions of the spiral nebulae, *Pop. Astron.*, **24**, 111–12.

Trumpler, R. (1930) Preliminary results on the distances, dimensions and space distribution of open star clusters, *Lick Observ. Bull.*, **14**, 154–88.

Turner, H. (1911) From an Oxford note-book, *The Observatory*, **34**, 350–4.

Vaucoleurs, G. de (1961) *Astronomical photography: from the daguerreotype to the electron camera*. London : Faber & Faber.

Very, F. (1911) Are the white nebulae galaxies?, *Astron, Nachr.*, **189**, 441–54.

Very, F. (1912) On stellar and nebular distances, *Knowledge*, **35**, 329–32 & 373–6.

Vorkin, D. de (1977a) The origins of the Hertzsprung–Russell diagram, *Dudley Observatory Report*, No. 13, 61–77.

Vorkin, D. de (1977b) W. W. Campbell's spectroscopic study of the Martian atmosphere, *Quart. J. Roy. Astron. Soc.*, **18**, 37–53.

Vorkin, D. de & Philip, A. (1977) eds. In memory of Henry Norris Russell, *Dudley Observatory Report*, No. 13.

Waters, S. (1873) The distribution of the clusters and nebulae, *Mon. Notic. Roy. Astron. Soc.*, **33**, 558–9.

Webb, T. (1863) Clusters and nebulae – the surface of the moon, *The Intellectual Observer*, **4**, 56–64.

Weyl, H. (1923) Zur allgemeinen Relativitätstheorie, *Phys. Zeitschr.*, **24**, 230–2.

Whitney, C. (1972) *The discovery of our Galaxy*, London: Angus & Robertson.

Whitrow, G. (1972) Hubble, Edwin Powell, *Dict. Sci. Biogr.*, **6**, 528–33.

Whitrow, G. (1978) Theoretical cosmology in the twentieth century, *Proceedings of the XVth International Congress of the History of Science*, edited by E. Forbes, 576–93. Edinburgh University Press.

Wilson, R. (1915) Recent observations of the nebulae and their bearing upon the problem of stellar evolution, *Pop. Astron.*, **23**, 553–62.

Wilson, R. (1923) The proper-motions and mean parallax of the Cepheid variables, *Astron. J.*, **35**, 35–44.

Bibliography

Wirtz, C. (1918) Über die Bewegungen der Nebelflecke, *Astron. Nachr.*, **206**, 109–16.
Wirtz, C. (1921) Einiges zur Statistik der Radialbewegungen von Spiralnebeln und Kugelsternhaufen, *Astron. Nachr.*, **215**, 349–54.
Wirtz, C. (1924) De Sitters Kosmologie und die Radialbewegungen der Spiralnebel, *Astron. Nachr.*, **222**, 21–6.
Wolf, M. (1908) *Die Milchstrasse*. Leipzig : Barth.
Wolf, M. (1912*a*) Die Entfernung der Spiralnebel, *Astron. Nachr.*, **190**, 229–32.
Wolf, M. (1912*b*) Das Spektrum des Andromedanebels, *Sitzungsber. Heidelberg. Akad. Wiss.*, **3A**, A3.
Wolf, M. (1912*c*) Über die Spektrum einiger Spiralnebel, *Sitzungsber. Heidelberg. Akad. Wiss.*, **3A**, A15.
Wolf, M. (1914) Report of the work of the Heidelberg Observatory, *Vierteljahresschr. Astron. Ges.*, **49**, 151–63.
Wolf, M. (1923) Zwei neue Veränderliche, *Astron. Nachr.*, **217**, 475.
Wright, H. (1966) *Explorer of the universe, a biography of George Ellery Hale*. New York: Dutton.
Young, C. (1888) *A text-book of general astronomy for colleges and scientific schools*. Boston, U.S.A: Ginn & Co.
Young, R. & Harper, W. (1916) The solar motion as determined from the radial velocities of spiral nebulae, *J. Roy. Astron. Soc. Can.*, **10**, 134–5.
Zwicky, F. (1929*a*) On the redshift of spectral lines through interstellar space, *Proc. Nat. Acad. Sci.*, **15**, 773–9.
Zwicky, F. (1929*b*) On the redshift of spectral lines through interstellar space, *The Physical Review*, **33**, 1077.

Index